T0155804

IoT Machine Learning Applications in Telecom, Energy, and Agriculture

With Raspberry Pi and Arduino Using Python

Puneet Mathur

Apress®

IoT Machine Learning Applications in Telecom, Energy, and Agriculture

Puneet Mathur
Bangalore, Karnataka, India

ISBN-13 (pbk): 978-1-4842-5548-3   ISBN-13 (electronic): 978-1-4842-5549-0
https://doi.org/10.1007/978-1-4842-5549-0

Managing Director, Apress Media LLC: Welmoed Spahr
Acquisitions Editor: Celestin Suresh John
Development Editor: James Markham
Coordinating Editor: Aditee Mirashi

Cover designed by eStudioCalamar

Cover image designed by Freepik (www.freepik.com)

Distributed to the book trade worldwide by Springer Science+Business Media New York, 233 Spring Street, 6th Floor, New York, NY 10013. Phone 1-800-SPRINGER, fax (201) 348-4505, e-mail orders-ny@springer-sbm.com, or visit www.springeronline.com. Apress Media, LLC is a California LLC and the sole member (owner) is Springer Science + Business Media Finance Inc (SSBM Finance Inc). SSBM Finance Inc is a **Delaware** corporation.

For information on translations, please e-mail rights@apress.com, or visit www.apress.com/rights-permissions.

Apress titles may be purchased in bulk for academic, corporate, or promotional use. eBook versions and licenses are also available for most titles. For more information, reference our Print and eBook Bulk Sales web page at www.apress.com/bulk-sales.

Any source code or other supplementary material referenced by the author in this book is available to readers on GitHub via the book's product page, located at www.apress.com/978-1-4842-5548-3 For more detailed information, please visit www.apress.com/source-code.

Printed on acid-free paper

This book is dedicated to the Supreme Divine Mother.

Table of Contents

About the Author

Puneet Mathur is an author, AI consultant, and speaker with over 20 years of corporate IT industry experience. He has risen from a programmer to a third-line manager working with multinationals like HP, IBM, and Dell at various levels. For several years he has been working as an AI consultant through his company, Boolbrite International, for clients around the globe, guiding and mentoring client teams stuck with AI and machine learning problems. He is a regular speaker at international conferences. He is also an Udemy Instructor with several courses on machine learning. His latest bestselling book, *Machine Learning Applications Using Python* (Apress), is for those machine learning professionals who want to advance their career by gaining experiential knowledge from an AI expert. His other hot books include *The Predictive Project Manager*, *The Predictive Program Manager*, *Prediction Secrets*, and *Good Money Bad Money*. You can read more about him on his website at www.PMAuthor.com.

About the Technical Reviewer

 **Abhishek Nandy** has a B.Tech in IT and is a constant learner. He is a Microsoft MVP for the Windows platform, an Intel Black Belt Developer, and an Intel Software Innovator; he has a keen interest in AI, IoT, and game development. He is currently serving as an application architect in an IT firm. He also consults on AI and IoT and does projects with AI, ML, and deep learning. He is also an AI trainer and drives the technical part of the Intel AI Student developer program at leading IITs in India. He has showcased a demo on Reinforcement Learning with Unity at SIGGRAPH 2018 in Vancouver, Canada. He has won four AI for PC Challenges at Intel. He was involved in the first Make in India initiative, where he was among the top 50 innovators and got trained in IIMA. He won an Early Innovation Game Dev grant from Intel; the game got published for Runs Great on Intel.

Link to DevMesh: https://devmesh.intel.com/users/abhishek-nandy

Link to his publication at the Intel Developer Zone: https://software.intel.com/en-us/user/78014

Link to Medium: https://medium.com/@abhishek.nandy81

Acknowledgments

I acknowledge the various engineers working at telecom companies, hi-tech agricultural farms, and in the energy sector who came forward to share information on how their operations needed further improvements and where AI could help; I used anonymous surveys conducted through personal and online means. I would also like to thank various company experts from multinationals like ABB, GE, and others who came forward to discuss topics related to this book, which needed their suggestions.

All the data in this book is anonymized and is not taken from any particular company, situation, or source. Any resemblance to actual data is only a coincidence. The datasets in this book are based on my experience working with clients and engineers; however, I have taken care to not take any such data from them and it is completely clean of any plagiarism.

The instruments and sensors including the drone and energy devices are not sponsored by any company nor did I get any fee or any incentive of any kind to use one over the other. The choice of all devices and sensors used in this book is entirely based on my independent judgement and experience.

Introduction

In late January of 2019, after attending a conference on IoT, I met a potential client in Mumbai regarding the use of machine learning in a factory setting. He wanted to know if machine learning could be used in his environment and, if so, what kind of business benefit could he look forward to in its implementation. I discussed a few use cases with him which involved the use of Industrial Internet of Things (IIoT). In business, there are two types of strategies: one is revenue growth and the other is cost reduction. If you are in a business where there is high revenue growth taking place, then you would not bother much about cost reduction. Your focus would be to expand your business. However, if you are seeing steady growth for a number of years and you have stiff competition, then the pressure on you is to not just to maintain your existing customers but also to reduce cost so that you can beat the competition. My client was in this mode of business. Most of the use cases I showed him were of revenue growth, which did not meet his expectation. However, the use case of doing an energy audit in his factory using IIoT caught his attention and he explained that he had a large electricity bill and he wanted me to implement an energy audit to help reduce cost. His next set of questions involved how much of cost reduction he could look forward to versus the investment that was needed to implement the solution. I gave him a small workout with a plan for its implementation, which was received well within his company, although we were doing this for the first time in a factory setting.

This book features one of those solutions, although not the complete one as it would need a separate book to do so. But the solutions in this book will help you get started in IoT and IIoT using machine learning.

CHAPTER 1

Getting Started: Necessary Software and Hardware

This chapter will introduce you to the world of single-board computers (SBCs). Many of you, hearing this term for the first time, may wonder what an SBC is and what it is used for. This chapter will explain SBCs and how they have developed historically. You will also learn about the most popular SBCs on the market such as Banana Pi, Raspberry Pi, and Arduino. In an in-depth comparison, I will explain the features of popular SBCs with regards to USB, storage, networking, and communication.

You will then learn about the Raspberry Pi and more specifically the Raspberry Pi 3 Model B+ because you are going to use it as a master node in the IoT and IIoT projects in this book. You will also find an in-depth explanation of the GPIO (general purpose input/output) pins located on the Raspberry Pi and their uses. You will then learn about single board micron rollers (SBMs), which are different from single-board computers. The single-board microcontrollers in IoT and IIoT applications are generally used as slaves to the single-board computers like Raspberry Pi. The most popular single-board microcontroller is the Arduino, and you will look at the types of SBMs in a detailed tabular format covering their processors, I/O modules, frequency, voltage, etc.

© Puneet Mathur 2020
P. Mathur, *IoT Machine Learning Applications in Telecom, Energy, and Agriculture*,
https://doi.org/10.1007/978-1-4842-5549-0_1

You will then look at Arduino Mega 2560 and its layout and learn about its GPIO pins. Next, you will learn about the most important topic of the book: IoT sensors and their types and applications. You will use some of them in the case study solutions. Also covered is the topic of drones because they can be used to collect data for a telecom application. Please note that flying drones requires a license in most countries and you should comply before trying to use one for any project. You will also learn about an Modbus device and how it is used to build a commercial application. Lastly, you will learn about the software programs that run all of these devices.

Note   Python version 3.x is used throughout the book. If you have an older version of Python, the code examples may not work. You need Python 3.x or later to be able to run them successfully.

Hardware Requirements

For running the exercises in this book you will need the following hardware:

- **Raspberry Pi 3 Model B+**

- **Arduino Mega 2560**

- **LCD/LED screen to output Raspberry Pi 3 Model B+**

- **Single phase energy meter: Modbus 220/230V, bidirectional, multi-function, RS485, pulse/Modbus output**

- **Drone: Industrial grade, heavy lifting which meets these specifications:**

  - **Max payload: 1.5 kg (3.5 pounds)**

  - **Range: 5 km**

  - **Time to fly: At least 30 mins**

Single-Board Computer

A single-board computer is a completely functional computer set on a single printed circuit board. What makes this type of computer unique is that the entire input, output, and processing such as graphics and numeric calculations all happen on the same board. While a SBC can be built as a high computing server, a more popular version of this type of computer is a small and compact machine.

Historically, SBCs were built to be educational and compact; now they are used in mainstream commercial applications.

Single-board computers are built on different microprocessors but they are all of a simple design. They are built to be handy and compact. Running a fast computer that occupies very little space and is stable and also portable adds to the charm of owning these machines.

One of the early implementations (May, 1976) of single-board computer was called the Diana micro; it was based on the Intel C8080A processor and was a very popular home computer as part of the BigBook series of computers. With the expansion of the PC market, the SBC did not progress further until now, when the PC has almost been replaced by tablets, laptops, and mobiles. You can go to `https://en.wikipedia.org/wiki/Single-board_computer` for more information.

Some of the advantages of using SBCs are shown in Table 1-1.

Table 1-1. *SBC Features and Advantages*

Features	Advantage
Price	SBCs are available for under $30 each today.
Form factor	SBCs are very small, almost the size of a palmtop, which makes them portable and handy to use.
Operating system	Linux variants are most popular. Windows IoT
Architecture	Two variants: slot support and no slot support.
Power efficiency	High

3

The main reason for the popularity of single-board computers is the price because they are available for under $30 each. Another perk is their compact and handy form factor; they fit the palm of the hand. There are variants that run Linux or Windows IoT or other operating systems. Most of the SBCs do not come with slot support; however, some do to support putting industrial grade cards on them. These small machines are highly efficient in terms of power and can run on home power single phase connections efficiently. Unless you want to plug them into a big LED or LCD screen, they do not need a separate power supply for the screen. They have their own 5-inch or 7-inch LED/LCD flat screens, which can be powered off the SBC itself. There is a flexibility of covers where the single boards can be put inside. From transparent ones to stylish cases to housing for clusters of 4 or 7 SBCs, there is a wide range of options to choose from for today's SBCs. This availability has extended the appeal of these computers by giving the user the DIY sense of achievement. From choosing the case to installing the operating system to using other types of software, everything is flexible and there are many options to choose from. Now let's look at the most popular SBCs on the market and their development.

Popular Single-Board Computers on the Market

Table 1-2 lists some of the most popular SBCs on the market, per https://en.wikipedia.org/wiki/Comparison_of_single-board_computers#Operating_system.

Table 1-2. Popular Single-Board Computers

Name	PCIe	USB[2]			Storage			Networking		Communication			Generic I/O	
		2	3	Device	On-board	Flash slots	SATA	Eth.	Wi-Fi	Bt.	I²C	SPI	GPIO	Analog
Raspberry PiZero W	No	No	No	OTG	No	microSD	No	No	b/g/n	4.1 + BLE	Yes	Yes	17	No
Raspberry PiZero	No	No	No	OTG	No	microSD	No	No	No	No	Yes	Yes	17	No
Raspberry Pi Model B+	No	4	No	No	No	microSD	No	10/100	No	No	Yes	Yes	17	No
Raspberry Pi Model B	No	2	No	No	No	SD	No	10/100	No	No	Yes	Yes	8	No
Raspberry Pi Model A	No	1	No	No	No	SD	No	No	No	No	Yes	Yes	8	No
Raspberry Pi 3 Model B	No	4	No	No	No	microSD	No	10/100	b/g/n	4.1	Yes	Yes	17	No

(continued)

5

Table 1-2. (*continued*)

| Name | PCIe | USB[2] | | Storage | | | Networking | | Communication | | | Generic I/O | |
		Device	2	3	On-board	Flash slots	SATA	Eth.	Wi-Fi	Bt.	I²C	SPI	GPIO	Analog
Raspberry Pi 2 Model B	No	No	No	4	No	microSD	No	10/100	No	No	Yes	Yes	17	No
Intel Galileo Gen 2[46]	1	Yes	No	1	8 MB Flash + 8 KB EEPROM	SD	No	10/100	No	No	Yes	Yes	20	12-bit ADC, 6 PWM
Banana Pi M3[137]	No	OTG	No	2	8GB eMMC	microSD	USB to SATA 2.0 adapter	GbE	a/b/ g/n	4	Yes	Yes	40	12-Bit-ADC (CON1 for Touch)

Name	PCIe	USB[2]			Storage		Networking				Communication		Generic I/O	
		2	3	Device	On-board	Flash slots	SATA	Eth.	Wi-Fi	Bt.	I²C	SPI	GPIO	Analog
Banana Pi M2	No	2	No	OTG	No	microSD	No	GbE	a/b/g/n	No	Yes	Yes	40	12-Bit-ADC (CON1 for Touch)
Banana Pi[11]	No	2	No	OTG	No	SD	SATA 2.0	GbE	No	No	Yes	Yes	26	12-Bit-ADC (CON1 for Touch)

7

As of this writing, the most popular SBCs today are the Raspberry Pi and Banana Pi. They are equally popular with hobbyists and serious users. While Raspberry Pi runs OSes like Noobs, Raspbian, and Windows IoT, Banana Pi run Linux and Android 4.x. Raspberry Pi developed from Model Zero to the current one, which is 3 Model B+ and, as you can see from Table 1-2, there have been huge hardware improvements. Raspberry Pi 3 B+ how has an ARM Cortex-A53 1.4GHz CPU, which is an improvement over the Raspberry Pi 3 B, which had an ARM Cortex-A53 1.2GHz CPU. The RAM size has not changed; however, there is an enhancement in Wi-Fi, which now has the capability of up 5GHz transfers. The Ethernet support has been increased from 100Mbps to 300Mbps. The use of Banana Pi is preferred by users who have projects that are closely linked with mobile applications since the operating system is Android which, although built on Linux, has Android as the kernel for its operations. For the projects in this book I have chosen Raspberry Pi. The operating system we are going to use is Raspbian, which is an adaptation of the Debian OS. We are not using Noobs because we want to make commercial-grade applications on the SBC and we are not using Windows IoT since we need an OS that has a desktop and development IDE embedded in it. So the choice of Raspbian is pretty obvious.

Use Table 1-3 to help you decide which Raspberry Pi to use for your projects.

Table 1-3. Raspberry Pi Technical Specifications

Raspberry Pi Platform	RAM	Processor	USB	Ethernet	Wi-Fi	Bluetooth	HDMI	Other Video	MicroSD
Raspberry Pi A+	512MB	700 MHz ARM11	1 port	-	-	-	Yes	DSI, Composite	Yes
Raspberry Pi B+	512MB	700 MHz ARM11	4 ports	10/100Mbps	-	-	Yes	DSI, Composite	Yes
Raspberry Pi 2 B	1GB	900 MHz Quad-Core ARM Cortex-A7	4 ports	10/100Mbps	-	-	Yes	DSI, Composite	Yes
Raspberry Pi 3 B	1GB	1.2 GHz, Quad-Core 64-bit ARM Cortex A53	4 ports	10/100Mbps	802.11n	4.1	Yes	DSI, Composite	Yes

(continued)

9

Table 1-3. (*continued*)

Raspberry Pi Platform	RAM	Processor	USB	Ethernet	Wi-Fi	Bluetooth	HDMI	Other Video	MicroSD
Raspberry Pi 3 B+	1GB	1.4 GHz 64-bit ARM Cortex A53	4 ports	300/Mbps/ PoE	802.11ac	4.2	Yes	DSI, Composite	Yes
Raspberry Pi Zero	512MB	1 GHz single-core ARM11	1 micro USB	-	-	-	Mini-HDMI	-	Yes
Raspberry Pi Zero Wireless	512MB	1 GHz single-core ARM11	1 micro USB	-	802.11n	4.1	Mini-HDMI	-	Yes

Before we move on to the topic of microcontrollers in the next section, let's look at a very important aspect of the Raspberry Pi architecture, which is the overlay of GPIO pins on the SBC board. These pins are used for communication with auxiliary hardware such as sensors or other microcontrollers that can do serial communication such as Arduino. Figure 1-1 shows the GPIO overlay of Raspberry Pi 3 Model B+.

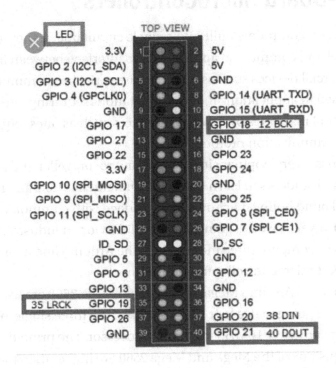

Figure 1-1. *GPIO pin overlay for Raspberry Pi 3 Model B+*

In Figure 1-1, you can see that there are a total 40 GPIO pins. They have a structure and it is important that you understand it in order to use it. The pins are numbered from 1 to 40 in the diagram. Pin 1 and 17 are for supplying power output to your device (3.5 volts). Pins 2 and 4 are used for giving a power output of 5 volts each. Pins numbered 6, 9, 14, 20, 25, 30, 34 and 39 are used for grounding the circuit. The rest of the pins are used for GPIO. You will be using this information later when you create a complete circuit for some IoT-based solutions.

Let's now discuss microcontrollers and how they are used.

Single-Board Microcontrollers

SBMs are microcontrollers built into a single circuit board; they are used in industrial and commercial applications to interface between industrial and commercial devices such as ones that use serial bus communication. They are used in applications to develop solutions requiring interfacing with industrial machines or network interfaces such as ones requiring Modbus communication protocol.

Arduino is a very popular single-board microcontroller and is used by hobbyists and students to learn hardware implementations and how to control and build hobby machines that interface with common C and C++ programs. However, we are going to use Arduino for an industrial-grade purpose. Let's compare three models from the Arduino line in order to select the best microcontroller for our purpose.

Arduino Uno, Arduino Mega, and Arduino Mega 2560 are the three models we are going to use for our comparison. All three single-board microcontrollers use a 16MHz frequency processor. The printed circuit board dimensions of the Mega and Mega 2560 are higher than the Arduino Uno, which is smaller than both of them by half in length at 2.7 inches x 2.1 inches to 4 inches x 2.1 inches. The Arduino Mega 2560 has the highest flash memory of 256kb, whereas Arduino Uno has the lowest flash memory of 32kb. Arduino Mega has 128kb. Flash memory is very important as far as SBMs are concerned because whatever programs you write for controlling the hardware through your application has to be written to the flash memory first. If the flash memory is low, you cannot write an industrial-grade application on top of it. Both the Mega models have higher EEPROM and SRAM than the Arduino Uno; also, the Mega models have 54 GPIO pins while the Uno has just 14 pins. Why do we need more pins for our applications? We will be using our applications for communicating with serial Modbus communication interfaces which may need two or three devices connected at the same time. If the number of pins is less, we will be limited in having parallel devices. So we need higher pins for this purpose.

So the best model for us is the Arduino Mega 2560, which has the highest flash memory.

Arduino Mega 2560

Figure 1-2 shows the GPIO overlay of the Arduino Mega 2560 and its communication architecture.

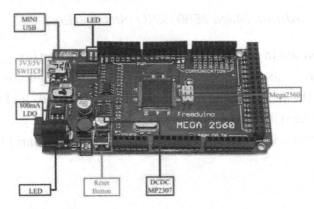

Figure 1-2. *Arduino Mega 2560 architecture*

You can see the area marked as GPIO pins in the single-board computer diagram in Figure 1-2. This model supports both digital pins and analog pins. Digital pins are used to control and communicate with devices such as digital sensors, and analog pins are used to communicate with analog devices such as analog sensors. This model has a USB interface, which is used for serial bus communication with another device such as a SBC like the Raspberry Pi. You will look at this in detail later in the part of the book on how to connect and communicate between the Raspberry Pi and Arduino. Figure 1-3 shows the overlay of GPIO pins on the Arduino Mega 2560 board.

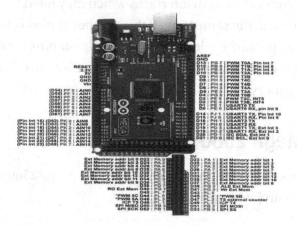

Figure 1-3. *Arduino Mega 2560 GPIO pins overlay*

As you can see in the figure, the pin numbers starting from D mean digital pins; these are the slots where you will connect digital sensors and devices. The pins starting from A are analog pins and can be used to communicate with analog devices. For now, this is what you need to understand. You will look at all of this in detail once you start to assemble your system.

IoT sensors

IoT is an extension of internet connectivity to everyday physical devices and objects. When devices like sensors and internet connectivity are added to it and embedded to physical devices like art objects or any other object that display information on creation date, an artist description of the object, recently modified date, etc., then the IoT is formed. We now have the concept of automatic reordering cabinets which can order food online as they gets empty upon consumption. By embedding IoT sensors in everyday objects we are making them smart and accessible for human betterment.

The common IoT sensors available on the market work with both the Raspberry Pi and Arduino because they use GPIO pins to communicate. Table 1-4 lists the common types of sensors that can be used both on Raspberry Pi and Arduino.

Table 1-4. *Common Types of IoT Sensors and Their Purposes*

Types of Sensors	Purpose
Machine vision	Optics and ambient light detection
Proximity and location	GPS location and presence of objects
Temperature	Detecting atmospheric temperature in air, soil, and water
Humidity/moisture	Detecting atmospheric humidity in air and soil
Acoustic	Detecting infra and ultrasound vibrations in the atmosphere
Chemicals	Detecting gas content in air, soil, and water
Flow	Detecting air and water flow in enclosed areas such as pipes
Electromagnetic	Detecting electromagnetic levels in the environment
Acceleration	Detecting the tilt of a connected electronic device
Load/weight	Detecting change in load or weight in the environment which is being monitored

You can see from Table 1-4 that there are a variety of sensors available and they range from digital to analog to measure and detect objects using Wi-Fi cameras to detecting a change in weight or load on an electronic device. The uses and applications of these sensors are limitless and can give fresh life to an IoT application. You will be using these common types of sensors in this book and I will outline them in more detail as you build your solutions.

Drones

A drone is defined by Wikipedia (https://simple.wikipedia.org/
wiki/Unmanned_aerial_vehicle) as an unmanned aerial vehicle
operated through remote control devices. Drones usually have a small
microcontroller embedded inside them which has the capability to detect
changes in air pressure, the proximity of objects, acceleration, etc. It has
many IoT sensors embedded within it, making it smarter than manned
planes. However, there are very few automatic drones that do not require
human supervision to fly and work. Most drones have human supervisors
who control their activities through a remote control mechanism.

You will be using an industrial grade drone in the solution for the
telecom domain case study. You will need an industrial grade drone that
can take payloads up to 3.5 pounds or 1.5 kgs and has a range of 5 km since
you will be flying it for detection purposes around an area in a particular
locality. The minimum flight time you need is 30 minutes because this is
the minimum time required to gather some decent data to apply machine
learning. These types of drones are definitely not cheap; they start at
$2,500.

Modbus Device

You will be using another device in the energy segment of the case study.
A Modbus device is a standard communication protocol for connecting to
industrial devices, such as machines to a computer. They are also used to
gather and send data. There is a master device or microcontroller that uses
the Modbus protocol to communicate with its slaves, which can be up to
247 in number. This makes the protocol very robust, which is the reason I
chose an Arduino Mega 2560 single-board microcontroller (because it has
a lot of space for analog and digital pins on its board for communication).

Disclaimer   Before proceeding further, I would like to warn you of the risk associated with using this Modbus device as the energy meter. It is an electric appliance and carries the risk of short-circuiting if not connected properly. It can also be dangerous to human life if not used as per its instruction manual. It can also burn and damage your Raspberry Pi and Arduino boards if wrong connections are made. So if you are not comfortable with electric connections, I strongly advise you to not use this device or to get expert help from a local electrical technician to make the proper connections. Neither I nor the publisher can be held responsible in any way whatsoever for any kind of damages, either material or to life. User discretion is advised.

In the industrial world, three-phase Modbus energy meters are used, but this is difficult to replicate in a normal environment, so I recommend using a single-phase energy meter instead, which uses the Modbus RTU communication protocol. There are plenty of options available from companies such as Schneider Electric and others.

Required Software

For running the exercise in this book, you will need Python 3.x installed. I recommend you use the default Python installation that comes with the Raspbian OS for this purpose. Python is a simple distribution and it does not require any installation whatsoever, unlike Anaconda. All the coding exercises in this book work on this version of Python. The exercises and solutions in this book do not support Windows nor have they been tested on any version of Windows. Using the Raspbian OS is a must to make them work. Please follow the steps for installation that are given in the installation section of this book.

You will also be using the Arduino IDE to communicate between the Arduino and Raspberry Pi, so the Arduino is going to work as a slave to the Raspberry Pi in your solutions. This can be installed via the `apt-get` command, which you will see later in the installation part of the book.

Summary

This chapter covered SBCs and how they have developed historically. You learned about the most popular SBCs in the market, such as Banana Pi, Raspberry Pi, and Arduino. In an in-depth comparison, you explored the features of popular SBCs in terms of USB, storage, networking, and communication features. You also learned about the Raspberry Pi and more specifically the Raspberry Pi 3 Model B+ that you will be using to use as a master node in your IoT and IIoT projects later in this book. You learned about the GPIO pins located on the Raspberry Pi and their uses. You also learned about the single-board microcontrollers and how they are different from the single-board computers. You also learned that the single-board microcontrollers in IoT and IIoT applications are generally used as slaves to the single-board computers like Raspberry Pi. You also saw that the most popular single-board microcontroller is the Arduino and you looked at a table about the types of SBMs and feature information like processors, I/O modules, frequency, voltage, etc.

You then looked at the Arduino Mega 2560 and its layout and learned about its GPIO pins. Then you learned about the most important topic of the book: IoT sensors and their types and applications. You also learned about drones, as they will be used to collect data for the telecom application in this book. You learned about a Modbus device and how it is used to build a commercial application. Lastly, you got a list of software that will be used through this book.

CHAPTER 2

Overview of IoT and IIoT

You looked at the definition of IoT in Chapter 1. In this chapter, I want to expand on it. What makes the concept of the Internet of Things work are the twin abilities of using small scale sensors and motors, and controlling their input and output through programming interfaces. The crux of IoT is the ability to remotely monitor and control the devices. This can be achieved through small programmable microcontrollers or microcomputers that interface with such devices. In using IoT, various technologies have come together like embedded systems, control systems, and digital and analog sensors. The concept is to embed sensors into everyday objects and make them smarter by monitoring and controlling them. However, not every IoT-enabled device catches the fancy of the consumer or meets a business needs.

A Closer Look at the IoT

There have been experiments with everyday objects such as smart choice, which monitors how much time you sit on it and monitors your vital statistics such as blood pressure, etc. In some of cases, the consumer does not see benefit buying these products.

© Puneet Mathur 2020
P. Mathur, *IoT Machine Learning Applications in Telecom, Energy, and Agriculture*,
https://doi.org/10.1007/978-1-4842-5549-0_2

Not all smart objects provide substantial benefits to the consumer or she may fail to perceive it as a substantial benefit for the extra price. In such cases, the smart product will fail at its launch. Although the technical capability does exist to make the object smarter, the consumer perception does not meet the expectations of the consumer.

There's also the concept of a smart home, in which objects like dishwashers are controlled out of a central monitoring system. However, due to its expense, this hasn't gained popularity.

Figure 2-1 illustrates the technological advancement in the field of IoT. It's reiterated from my book *Machine Learning Applications Using Python.*

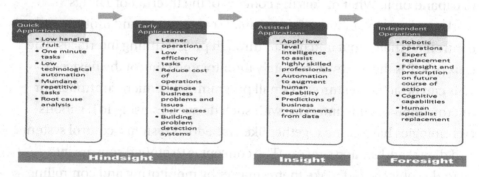

Figure 2-1. *Machine learning technology adoption process*

The IoT is presently in the stage of quick applications and early applications. You know that track applications stage it is trying to forget the past such as switching all end of the home electrical appliances. Applications stage today's benefit by reducing the cost of monitor dinosaurs installing IoT sensors and cameras for security etc. The next stage of applications sends it to you state applications stage. By using the Internet of Things, the effort will be to augment the capability to predict the needs of humans, such as ordering groceries inside smart cabinets by using computer vision cameras to monitor the number of common household items inside it and predict when they will need to be replenished. Such applications make it to the third stage, the assisted applications stage, or the fourth stage, where robotic operations augment human capabilities. The operations that a human servant would perform at home today will be performed by a robot, such as cleaning utensils and clothes. Such robots would have the ability to predict any future problems that can happen in the household and also prescribe remedies for them. For example, a reordering robot would check if any of the items are getting low and then send a reorder online to a nearby store that offering a discount plus home delivery. It could also wait for confirmation of the placed order. If it does not get confirmation in the specified time, it would send an alert to its human master via SMS or even walking up to them and showing the problem to them directly. It could also suggest alternatives such as ordering the product from a faraway store that does not do home delivery. Such a robot would definitely be an asset to any household. This is just an example of how robotic operations could become predictive but also prescriptive in suggesting an alternate solution to a potential problem before it arises. Table 2-1 shows some example applications of the IoT in the current scenario.

Table 2-1. Applications of the IoT

Technology Phase	Application of IoT
Quick applications	Home automation
Early applications	Personal wearables
	Smart city
Assisted applications	Farm automation
	Retail automation
	Smart car
Independent operations	Medical surgery

You can see home automation move to the applications stage in the table above because people are currently automating mundane home tasks that they are performing repeatedly, day in and day out, such as switching off appliances and monitoring and controlling them through personal phones like Android and iPhone devices. These applications allow you to control devices like smartwatches, which can accurately gather personal bio information like number of steps taken, heartbeats per minute, blood pressure rate, and oxygen carrying capacity. These apps can generate reports on a weekly, monthly, or yearly basis and use that data to show your health routine.

Another development in the IoT arena is that of the smart city, which is at the stage of early application. The most prominent one is Singapore, which is trying to truly become one in every way. Smart buildings are being built so that they take in maximum sunlight and are self-dependent by using energy recycling methods such as rainwater harvesting solar panels for producing off-grid energy for the residents. A smart city recycles all its waste and consumes and recycles all its resources such as water, oil, garbage, and gas.

Farm automation is one of the examples of an early application of automated operations now possible due to the IoT; however, it has to move to the predictive stage where it is able to produce the crop cycles such as crop failure or crop success. The Weather Company by IBM (`www.ibm.com/weather`) is trying to change the way farming automation is being done by trying to predict the weather for crops. Today's farming is completely dependent upon weather cycles and seasons. Although everything in relation to farming, from sowing the seeds to harvesting them, has been automated, the ability to predict the weather for farming will truly move it to the stage of independent operations.

Smart cars are another such application. When you can drive a smart car in crowded areas using IoT sensors such as computer vision sensors and GPS locators, this application will be in the stage of independent operations; also, when it can predict and prevent accidents.

Medical surgery has to a certain extent attained the independent operation level, like medical surgery for cancer where the robot can independently do surgery by augmenting the surgeon's tasks. 2019 game search applications Other surgeries will mature over time as the world does more research on how to add surgery skills to robots so that they become more efficient.

These are some of the examples of applications of for the IoT to give you a context on how they are placed with technology adoption.

Commercial Uses of the IoT

Let's now look at some of the successful IoT applications that are working well in the market.

A 2019 survey by teknowlogy (`www.device-insight.com/wp-content/uploads/2019/03/the_iot_survey_2019_highlights_device_insight.pdf`), was carried out on IoT vendors based on parameters such as user review matrix, effectiveness, efficiency, business value, solution timeline, visualization, support of new concepts, innovation, pricing model, product satisfaction, partner ecosystem, performance satisfaction, flexibility, ease of use, and customer experience. The companies covered are mid-sized ones and this is where the major implementations of IoT are happening. I've picked the best parameter, which is customer experience, and I'll discuss the solutions that these high-scoring companies have provided. The reason for choosing this parameter is because even though you may have the best products as per the rest of the parameters like innovation and visualization, if your customer is not satisfied, you can't sustain that product in the long run. Table 2-2 shows the three companies with great IoT solutions that are being used commercially today.

Table 2-2. *Top Three solutions of 2019*

Company	Top Solution	Applications
C3 IoT	C3 AI Appliance	The C3 AI Appliance™ powered by Intel is for organizations that need to deploy artificial intelligence (AI) applications to analyze massive amounts of data without compromising stringent data governance, compliance, and security requirements. `https://c3.ai/products/ai-appliance/`
Siemens IoT EMEA	MindSphere	MindSphere is the cloud-based IoT open operating system from Siemens. It connects your products, plants, systems, and machines, enabling you to harness the wealth of your data with advanced analytics. In addition, it gives you access to a growing number of apps and a dynamic development platform as a service (PaaS). MindSphere works with all popular web browsers. `https://siemens.mindsphere.io/en`
ARM Mbed	Mbed OS 5	Mbed OS is the leading open-source RTOS for the Internet of Things, speeding up the creation and deployment of IoT devices based on Arm processors. With the Mbed OS, you can develop IoT software in C++ with the free online IDE, generate optimized code with the Arm C/C++ Compiler, and run it on hundreds of hardware platforms. The Mbed OS stack includes TLS, networking, storage, and drivers, and is enhanced by thousands of code examples and libraries. `https://os.mbed.com/`

The three solutions that I picked to showcase here work well commercially. The top solution according to the survey is C3 AI Appliance. Appliances that run on different servers are very hard to develop, and it's harder to meet all customer expectations. Given that this is an AI platform-based appliance running on an Intel server which allows analyzing massive amounts of data and also meets the data governance, compliance, and security protocols, it's amazing that the customer experience is very high. The customer testimonials on the website are very positive so it's no wonder it gets a top slot in the survey.

The next product picked from the survey is by Siemens (`https://new.siemens.com/global/en/products/software/mindsphere.html`). It's a cloud-based IoT open operating system that connects products, plants, systems, and machines, and lets the customers harness it with advanced analytics. It is an excellent example of an implementation of a platform as a service (PaaS). It's known as MindSphere. In order to make this another amazing product, given that it is a platform as a service, MindSphere truly works from all popular web browsers such as Chrome, Mozilla, and Firefox, to just name few.

The third product is from all Mbed (`https://os.mbed.com/`), which created an operating system open source RTOS for IoT, speeding up the creation and deployment of IoT devices based on Arm processors. The main development languages used by this OS are C and C++. It runs on hundreds of hardware platforms such as Raspberry Pi and also has a community that is very active. This platform has various drivers and code examples in libraries available for rapid implementation of all the three functions to compete with each other especially in the MindSphere IoT operating system from Siemens because they are the core for delivering total solutions. C3 AI Appliance takes the concept of an OS further by its ability to host on a private cloud; this solution is boon for those companies that want complete control of their IoT solutions. For companies that want open source solutions, the MindSphere solution by Siemens can work on hardware platforms and specifically uses C and C++ code for software solutions.

You have seen three different companies and three different solutions, and the excellent part is they are delighting customers with their experiences.

IoT Trends for the Future

It is important to look at the top trends that are emerging in the field of IoT in order to get a holistic picture about it. A lot of surveys and prediction of all the IoT trends for the year 2019 have been done. According to ZDNet story on 2019 IoT trends by Eileen Brown (`www.zdnet.com/article/whats-next-for-2019-iot-trends-and-predictions/`) a Northstar survey of global consumers showed that intelligent homes will become mainstream and the next hot thing will be interdelivery options to consumers that will be delivered through smartphones and GPS positioning data. It also predicted better quality in healthcare due to the deployment of connected sensors in hospitals to ensure that the time to find critical medical equipment is reduced. The survey also talked about the function of smart cities by delivery cost reduction benefits in terms of better waste management and citizen engagement for revenue stream opportunities and energy efficient buildings. This has been in the mainstream now with cities like Singapore, which has already shown that a smart building can reuse its waste including water and garbage resources to be self-sufficient.

A survey by GoodWorkLabs on the IoT, published in February 2019 (`www.goodworklabs.com/iot-trends-2019/`), talked about edge computing, which is distributed computing performed on distributed smart edge devices instead of in a centralized environment for indoor areas or industries where there is no requirement for a centralized network for processing. The survey lists as an important item security for IoT devices and notes the efforts taken by hardware manufacturers like Siemens and GE, which are making smart devices that focus on the endpoint security of the user. The survey also talks about its application in the healthcare and manufacturing industries. The smart beacon RFID tags are part of the new industrial revolution that is going to take place on devices by 2020. The governance of consumer IoT industry is also pointed out to be a trend that will advance the development of smart homes. Another major development is that large players in the consumer industry are coming together to form subscription offerings. This will happen in the areas of utilities, food companies, and service aggregators. There are talks about the growing market of connected smart cars which will be accessible through the smart apps on your mobile phones, which will show you real-time diagnostic information about your car using IoT technology. All the sensors residing within your car will give you data about your current location but also what is going in your car such as tire pressure. The survey welcomes the beginning of 5G in 2019, which will become the backbone of IoT technology by supporting the interconnectivity of IoT devices. The 5G network will allow smart devices to produce and send data in real time, which has not been imagined so far.

The website iotforall.com (`www.iotforall.com/top-iot-trends-rule-2019/`) reports that 2019 will be about big data and artificial intelligence. The topmost forecast by Gartner says that there will be a rise of 14.2 billion IoT deployments. It talks about connected clouds. Many companies will rely on clouds to store data because cloud storage is connected to bandwidth. Accessing data is the reason for choosing different cloud services, such as the offerings by giants like Amazon AWS Cloud Platform, Microsoft Azure, and the Google Cloud platform. Today's need is for these different clouds to speak to each other; this is known as connected clouds, such as the partnership formed by Oracle and Microsoft by connecting their cloud platform services. The survey also talks about connecting private and public cloud for any company for servicing their data storage needs. The survey also talks about edge computing, which is going to break through in 2019. The survey talks about digital twins, which are also known as hybrid or virtual prototyping businesses using special tools like AI and machine learning and IoT to improve their customer business experiences by streamlining their data operations. This survey highlights that 5G will be game-changing in the IoT market by bringing about revolutionary applications. One point mentioned by this survey is sensor innovation, although it does not talk about what type of sensors will be developed in the market, but it says that the new special purpose sensors will lead to efficient and effective use of power consumption using deep neural networks, leading to new age architectures and low power IoT and endpoint devices. The development of new algorithms using data from these sensor technologies will lead to new implementations in the domains where they are applied. The survey prediction for social IoT, which is giving rise to social anxiety and is going to transfer the business sector from consumer devices to large scale manufacturing, is like a warning for the IoT applications and solutions implementers in order to take social responsibility into considerations when building their products. The social aspect is highlighting the need for a human being to accept the current state of human beings in accepting IoT in their everyday life. Although the survey does not point to any specific issues, it says that as online gains acceptance, social, legal, and ethical issues will crop up.

A survey by Network World covers the top 10 IoT trends for 2019 and beyond (www.networkworld.com/article/3322517/a-critical-look-at-gartners-top-10-iot-trends.html). The first trend is artificial intelligence, which is not a big surprise. The second one is social, legal, and ethical issues. Info Linux and data booking which river is commercializing of Information and making the data available agents and submergence on subscription or condition basis which this Gartner study reports. IV is shift phone edge intelligent mesh. The edge mesh defined as a network of connected edge devices true internet connect with cloud end-user devices together at a high level known as edge mesh. The study points to the fact that instead of having such devices working in silos, this will get them connected to intelligent machines and will create a network of networks of their own. The trend, as seen by Gartner, is that of IoT governance, covering the legal and social aspects and ethics involved in the implementation of IoT applications.

The other oils like new user experiences and new Wireless Networking Technologies for IoT salwar significant thanks that the study mentions.

PCMag (www.pcmag.com/feature/365945/8-iot-trends-to-watch-in-2019/8) came out with its trends to watch in 2019. One issue that the IoT is currently facing is surcharge protection against ransomware. It mentions the case of the city of Atlanta, when city systems such as the water services system were hacked. by a hacker it systems ransom technicians that all the smart devices this threat mobile device Management Solutions have to give up in order to protect them against search attacks. An interesting case has been highlighted survey date of IoT Technology tablets used to keep food safer. Example of cold food storage operator which offers hybrid cloud to keep food safe IoT is helping companies teacher and humidity cold storage facilities.

Another issue highlighted through this article is the use of IoT to simplify maintenance in manufacturing. The use of sensors to gauge problems in manufacturing has been discussed where the actual site technician only thumbs the sensors on the touch screen to detect problems that could arise with the machine in the near future. Factory workers are going to have IoT-based wearables on their bodies to manage, monitor, and control them from control centers, as highlighted in this article.

A TechGenix article on trends and predictions for 2019 talks about two interesting aspects of the new IoT platform (http://techgenix.com/iot-trends-2019/). It talks about IoT platform vendors that are not focusing on use cases and integrity. The hyperscale clouds are going to be labeled as IoT destinations; a marriage between hyperscale cloud IoT platforms is going to emerge from this year onwards. Another trend is the development of the managed services market for the IoT. The emergence of IoT services will include management operation of IoT networks and other IoT assets. In the article, the prediction is that these offerings will be customized for smart products made for smart homes to support IT solutions.

The last survey is the article by Forbes (`www.forbes.com/sites/bernardmarr/2019/02/04/5-internet-of-things-trends-everyone-should-know-about/#752315be4b1f`) written by Bernard Marr in February 2019. Among other things, it predicts that devices will become more vocal. For example, Alexa, Siri, and Google Voice commands are going to see their emergence.

We have looked at money predictions the trains the field of IoT silent love and see what the common friend is within all of these. Summaries of these predictions and trends for IoT are shown in Table 2-3.

Table 2-3. *Common Predictions and Trends for IoT from 2019 Onwards*

IoT Trend Area	Description
Smart homes	Home automation in electrical appliances, cleaning services, etc.
Edge computing	Decentralized and distributed edge devices work in remote areas of operations.
Endpoint security	Security systems against ransomware for IoT devices.
5G service introduction	5G will enable IoT devices to communicate data with higher bandwidth and faster to centralized clouds.
Artificial intelligence	Machine learning and deep learning capabilities and robotics
Cloud services for IoT	All cloud services will now be offering IoT-based solutions.
IoT sensors innovation	Many sensors such as thought and voice sensors will be innovated.

After having looked at the common future trends and predictions in IoT from the best in the industry, I now present to you a trend that is definitely going to see its emergence starting in 2020. The Thought AI has yet to see its commercial implementation, including inventions of thought sensors. I will discuss all of this in the next section of this chapter.

A Closer Look at Thought AI

You have seen the best IoT commercial applications in the field of IoT. You have also seen the trends for the future as predicted by the aforementioned articles. However, as a technical visionary, I do not think that any of these trends are going to be as game-changing as a certain application. Yes, I am talking about something that is going to not only revolutionize but also change the way humans and machines interact with each other. It is going to create a world that is going to be more intuitive, easy going, and responsive to human needs. It will bring about a change as revolutionary as the invention of the PC or the mobile phone. It will see technology move up to a level that has not been seen before. It will make technology rise extremely close to the Homo sapiens as could ever be. It will be the age of machines as friends and rogues. Of course, such technology has the equal potential of being anti-human as well. I re-quote from my first book, *Machine Learning Applications Using Python*, "It is not the artificial intelligence but the human intelligence behind the artificial intelligence that is going to change the way we live our lives in the future". Please remember that this section is futuristic and that Thought AI is a basis for the near-future as it is yet realized. It can happen in the future. Also, you need to know that implementing Thought AI will be costly as it is in the research and development stage.

As a visionary, I present to you the technology for tomorrow. This is what I call Thought AI. If you look at the input methods that exist in any machine from a human being, there are just two. The first one is typewriting via a keyboard and the second one is voice dictation via a speaker. Table 2-4 lists the differences between the type and voice inputs to emulate machine-like robots on our computer.

Figure 2-2 shows the process that the human brain has to undergo to produce a typewritten work. You need to read them together to understand the advantages of Thought AI.

Table 2-4. *Advantage of Using Thought AI*

Typewriting	Voice Dictation	Thought AI
Advantages		
It is an old way.	It is a new way.	It is going to be the new way.
Many people know how to type on a keyboard.	Everyone who can speak can work with this input.	Everyone who can think can work with this input. Even persons with disability can use it.
Keyboards in new devices are now built into the software. No external device is necessary to type.	Microphones are built into every device these days including PCs, laptops, mobiles, and tablets.	Commercial grade thought sensors have yet to be developed. Once in the market, they will be built into every device we know of today such as mobiles, PCs, laptops, etc.
Private activity as long as the screen is not visible to others.	Not as private because what you speak is audible to everyone around you.	Totally private to you. Nobody except the device that is running the Though IoT sensor knows what you are thinking.

(*continued*)

Table 2-4. (*continued*)

Typewriting	Voice Dictation	Thought AI
Disadvantages		
Takes time	Takes time but it's faster than typing.	Advantage: Real-time input as the term will get popular "WHAT YOU THINK IS WHAT YOU GET."
Requires special practice to type fast and efficiently.	Does not work in noisy environments.	Advantage: Does not require any special skill like learning to bang a keyboard. You just think.
Tiresome activity if pages of input are needed. Hands get tired of typing.	Speaking continuously for large input can be tiring.	Advantage: You get tired only if you get tired of your thoughts.
Not real-time to our brain activity. There's a lag between our thought flow and typing activity on the keyboard. The brain has to do extra processing to convert thoughts to the motor activity of spelling each word and then giving each finger the command to type.	Lack of privacy because the voice is audible to everyone around the human being.	Advantage: Can't get more real-time because the inputs are received via the thought sensor as and when you think.
	Challenges of different accents around the world for just the English language.	Advantage: No issue with accents because thoughts are universal in a particular language.

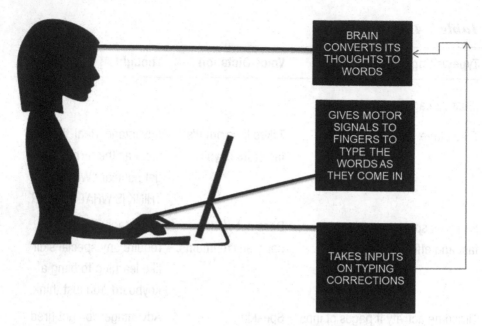

Figure 2-2. *The typing process for a human brain*

Typing

Typing is an old way of doing things. It's how laptops and PCS have been receiving inputs from human beings since the beginning. The advantage is that many people are familiar with how to type on a keyboard. However, for PCs and many laptops, it's necessary to have an external keyboard. Typing on a keyboard is a fairly private activity as long as the screen is not visible to others. In terms of love letters, screen privacy is important.

In terms of disadvantages, typewriting takes time and skill; in fact, special practice is required to type quickly and accurately. Proper finger placement on the keyboard requires practice. Regardless of skill, typing can get very tiring and can cause serious injury like carpal tunnel issues. The other disadvantage is that it is not a real-time input from the brain; the lag between our brain and any activity on the keyboard has to do with the extra processing of converting thoughts to the motor activity of spelling each word and then giving each finger the command to type appropriately. This can be seen in Figure 2-2 where I show you the process of a human typing.

Voice Dictation

Now let's look at voice dictation. It's a new way to import data into computers and other machines. The major advantage over typing is that anyone who can speak can work with this type of input. Another advantage is that microphones are in every device these days including PCs, laptops, mobiles, and tablets. One of the greatest disadvantages of technology is that it is not a private activity; anybody who is nearby can listen to what you are speaking to the machine. Another disadvantage of this type of input is that it takes time; although in comparison to typing, it is faster because the brain does not have to convert teach word to a finger movement. The biggest disadvantage of this type of input is that it does not work in a noisy environment; if some people talking at the top of their voices, the system may fail to detect your commands properly. Another disadvantage is that speaking continuously for a long time makes the human jaw tired because of the use of the muscles around the mouth. Another difficulty for voice typing apart from privacy is different accents; for example, there are so many different English, African, American, and Asian accents. This is the reason why voice dictation programs aren't very popular.

Love will come to technology the future this is my vision of how this technology is going to develop. It is going to be the new way of doing things. Its greatest advantage is that this type of input is available to everyone who can think. People with disabilities can use this type of input. The best example is the popular scientist Stephen Hawking, who had physical disabilities but could communicate with the machine through his parts. The current disadvantage is that commercial grade thought IoT sensors have yet to be developed. Once this type of IoT sensor is developed and on the market, they will be built into every device. The key advantage of Thought AI is that it provides total privacy; nobody else except the device reading your thoughts knows what you are thinking. This is why this type of input is going to be popular; it's going to be a real-time input from the brain. and the catchy phrases that is going to nice ok this concept is what you think is what you get the advantage that It does not require any special skill like learning to bang on a keyboard. Another advantage is that there is hardly any physical activity required for this kind of import; you only get tired if you get tired of your thoughts. The biggest advantage of this input is that there is hardly any lag between what you think and what will be available to the machine through the thought IoT sensor. There won't be any issue with accents because thoughts are universal and they don't have any accent, so they don't care if you speak in African English or Asian English or American English. So now you have looked at the futuristic Thought AI. It will truly revolutionize the way we will interact with machines and the way machines interact with us.

Thought AI Technology

Thought AI technology will give rise to an independent network. The "Thought Network" is going to be all-pervasive and all-prevailing as it is going to be a connected network that will have the ability to read public thoughts of the population of a country in minutes. Just as we have social media that is being harnessed by governments, so too will the "Thought Network" be harnessed. This will give rise to new media, the "Thought Media." Many new devices will need to be invented to cater to the Thought Network and the Thought Media.

For simple things like eating ice cream there would be dedicated stations that would read your thoughts and then shop for ice cream for you based on your mood and thought patterns. A wearable thought-enabled garment would analyze its wearer and look for patterns such as suicide patterns or obsession patterns and alert the person and their connected thought doctors. Special thought sensors could be put on repeat offenders in order to detect their crime-thinking patterns and prevent them from committing a crime. This is truly predicting a crime event before it gets converted to action from inside a criminal's head. It could also detect obsessive thinking patterns in people and try to prevent harmful actions by giving coaching or counseling sessions in those areas where the thought patterns were more prominent. Happiness patterns could be exploited by businesses by looking for obsessive patterns about certain goods and promoting them according to your obsessive needs.

Speed of thought would become the new currency in such a society where IoT sensors would be regulated especially around the public thought. A person would have the ability to dictate and customize the IoT device on which thoughts to keep private and which ones to keep public.

They would be the development as a thought identification as teach thought pattern of a person to identify that person through the machine train. Devices which would do not to paint wanted to print the pattern of your thoughts give you the ability to do that. Part 2 video occasion of this thought also thought to voice pair conversion of thoughts to voice well you to share it with other people instead of email d used for. Communication voice. The concept is of inner voice outer voice very similar to private thought and public thought into play.

All this can be achieved only if we have good thought IoT sensors with thought transmitters and thought receivers. Intellectual property rights would have to be changed in order to recognize people's thoughts as intellectual property. Thought-to-text would be recognized as a legal document. Such as Society of the future public trolls with thought readers gauge the mood of the people birthday coming very common. Of course, some people could opt for thought jammer devices to block public readers. Preventive measures based on a person's thought pattern and criminal tendencies would become commonplace in such a society. You could in such a thoughtful time get offers in real time based on your public thoughts; for example, your repeated thought pattern of pizza would get you new promotional offers related to it. There would be no text or written files, but thought files would be shared between the personal thought readers of people. Thoughts of the best would be admissible in a court of law as evidence. Doctors would get the ability to diagnose diseases based on thought reports. Retail stores could offer discounts based on your obsessive thoughts on certain objects. "Desire, think, and buy" products and "Desire, think, and sell" would become the order of the day. Dating in such a society would means sharing your private inner voice and all personal thoughts with your date's personal thought reader.

The biggest change it would usher in is that your thoughts are not private by law anymore as there is a machine that is continuously reading the voice in your head. Hacking this device, if not secured, would become a real possibility. The possibility of mass brainwashing would make this technology controversial. Thought hacking is one of the biggest problems this technology would face during the early stages.

Industrial Internet of Things

It is now time to understand the Industrial Internet of Things (IIoT). The IIoT plays the industrial equivalent of the IoT; it is computers, devices, and objects that may be connected to each other and share shop data with each other. The collected data resides as a central service like the cloud-based service or resides in private computers which they access for their own use. IIoT works in the manufacturing world. IIoT allows access to industrial data at better speeds and direct use of connected devices in the factory environment. Many IIoT protocols such as the Modbus or MQTT protocols allow IoT devices to directly link with industrial devices like assembly-line machines, processing machines, boilers, or heat exchangers, to just name a few. So what is the advantage of connecting today's factory machines, which are already monitored by their own software of some kind? Yes, it is true that any kind of an industrial machine that is built today has some kind of monitoring and control system which has data that is available from within the machine since it has a limited storage capacity recycled for continuous operation. Let's consider a heat exchanger, as part of a power plant. It has data about temperature details like heat and pressure within the control panel. However, if we want to make the single unit machine more efficient and productive, there is some kind of a pattern operating them and this data can be centralized on a public or private cloud. You could also apply some IoT sensors to the heat exchanger in order to collect different data that is not generated within it. So we see in this simple example that the biggest benefit can be cost reduction if we are able to collect data from machines currently sitting in silos and then apply learning and deep learning on them to gain an advantage. This can be seen from Figure 2-3.

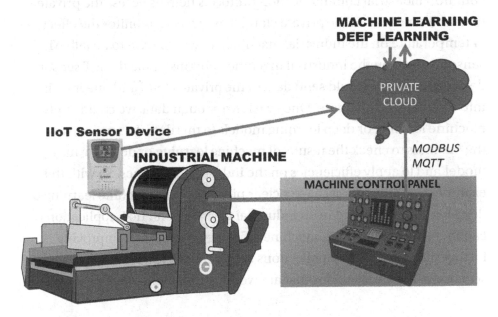

Figure 2-3. *How IIoT sensors and industrial communication protocols work together with machine learning*

In the diagram, there is an industrial machine with its own control panel. (It does not really matter what kind of industrial machine process; I am showing you that it's universally applicable.) There all are four major IoT systems. The first one is the industrial machine itself. The second is its hello. The third is the private cloud. The fourth are the IoT sensor devices that affect the industrial machine. V continent really is the heart of IoT system which provides the real benefit behind machine learning or deep learning model that is built using the private or public cloud. Industrial machines cannot communicate directly with IoT sensor devices, so they have a control panel that faces into the parts of the machine to monitor and create data. We generate machine-level internal data access through the control unit interfacing a device using the MQTT protocol. These common industrial communication protocols help us access the private data and send it all to the private cloud. If we were to monitor the effect of temperature on the industrial machine, we would have to install IoT sensor devices on the industrial machine. After installing the IoT sensor device, this device would send data to the private cloud and integrate it into a centralized database. Once we have enough data, we could apply machine learning or deep learning models to this data. It is now up to the business to check the results of machine learning or deep learning model and to apply efficiencies on the industrial machines. So with this example, you should now have a clear picture of how machine learning or deep learning is applied in an industrial environment. IIoT applications can include energy audits, machine efficiency checks, and improving factory efficiencies. The applications are limitless. Now let's look at some successful IIoT applications that are working well in the market.

Commercial Uses of IIoT

Let's look at some of the successful IoT applications that are working well in the market. There are not many players who focus on this market. Nevertheless, we look at a few who are dominating this space. An IIoT player known as Corlina (`https://corlina.com/product/`) has a product for the market that offers edge device security to make the devices visible and to certify them as trustworthy. It provides the first line of defense against security attacks to IIoT devices. It has a Smart Factory solution which allows factory owners to monitor and control IIoT infrastructure from inside the factory or remote location allowing even thousands of devices operating on the shop floor.

Next is the giant conglomerate General Electric. Predix (`www.ge.com/digital/lp/predix-industrial-internet-platform-brief`) is an IIoT platform in which real-time factory originated industrial data can be put into actionable insights and prediction models. It uses data from edge devices and helps the industrial owners to harness this digital industrial data into predictive models using digital twins and industrial AI.

Schieder Electric offers Wonderware (`https://sw.aveva.com/monitor-and-control/industrial-information-management/insight`), an IIoT-based systems and solution for energy management and industrial automation. Industrial data is collected in a fast and secure manner to provide device integrations such as edge devices, HMI process visualization software, and monitor-and-control solutions for factories. It also has a solution named Edge to Enterprise where it uses data from edge devices, moves the data to the cloud from the factory settings, and uses machine learning and AI to derive meaningful business actions out of it.

IoT and IIoT Differences

Thus far, you have looked at IoT and IIoT in detail. You now understand the commercial developments that are happening in these areas. However, it is important to know the difference between them; this is a very common question that I get asked in conferences and seminars. Table 2-5 highlights these differences.

Table 2-5. *Differences Between IoT and IIoT Applications*

Parameter	IoT	IIoT	Description
Focus	Consumer	Industrial	IoT caters to the consumer at large while IIoT focuses on industries and factory settings.
Accuracy and precision	Low	High	Accuracy and precision for IIoT applications is higher than in IoT applications because industries need to have higher fault tolerant systems because they deal with giant machines.
Risk impact	Low	High	IIoT systems work in spaces such as aerospace, healthcare, etc. where the room for error is very low so the risk impact is very high in comparison to consumer-based IoT applications.

The focus of IoT applications is the consumer at large and the focus of IIoT is industrial applications. This actually determines the ground for the next two parameters: accuracy/precision and risk impact. The accuracy and precision of industrial grade applications should be higher because they sometimes deal with hazardous processes and impact many lives on the factory shop floor. An error by an IIoT application could cause a company millions or even billions of dollars of losses and endanger the human lives involved in the production processes. This is the reason why IIoT devices use sensors that have higher accuracy levels than those used in personal IoT applications. The risk impact of an IIoT application failure is very high because many lives and a lot of business money and resources are at stake. IoT applications are also used in critical processes like healthcare where lives could be at danger; however, they are of personal nature. Nevertheless, every life is important and has value. With this, we come to the end of this chapter. Next you'll see how machine learning is used with IoT and IIoT in Python with example code.

Summary

In this chapter, you learned what IoT and IIoT are all about. You now understand the differences in their implementation based on focus, accuracy, and risk impact. You looked at some commercial IoT applications such as C3 AI Appliance, MindSphere (the cloud-based IoT open operating system from Siemens), and Mbed OS (the leading open-source RTOS for the Internet of Things). Next, you looked at the transformation of AI for human needs from typewriting to dictation to Thought AI, which is in the future. You looked in depth into IIoT and how industrial communication protocols work together for industrial applications. You explored the commercial use of IIoT by companies such as Corlina, which certifies edge devices and the General Electric Predix platform for IIoT devices and solutions, to name a few. Lastly, you learned the differences between IoT and IIoT in order to understand how they are implemented in real-world environments.

Summary

CHAPTER 3

Using Machine Learning with IoT and IIoT in Python

The purpose of this chapter is not to get you going on the hardware and the software but to show you how the interconnection between the hardware and software happens in the IoT. You'll get an introduction here to this concept and then later in Chapter 5 you'll learn how to do the hardware and software setup when I show you how to install Raspberry Pi, Arduino, and other devices from scratch.

This chapter is designed to show how the entire system is tested and made ready to implement a machine learning model. I will show you step by step how to use the hardware and software to get data from IoT sensors and store that data in flat files. After this, you will do machine learning processes, especially my seven-step machine learning lifecycle process from my book *Machine Learning Applications using Python*, in Chapter 1. Figure 3-1 shows a block system diagram of how the communication between hardware and software components is going to happen in this chapter.

© Puneet Mathur 2020
P. Mathur, *IoT Machine Learning Applications in Telecom, Energy, and Agriculture*,
https://doi.org/10.1007/978-1-4842-5549-0_3

Testing the Raspberry Pi Using Python

The first step to creating a solution using any hardware is to test the hardware-software interconnect and see if it is working fine. Figure 3-1 shows the system diagram I built in order to test the hardware and software along with the IoT sensor data by applying machine learning on it.

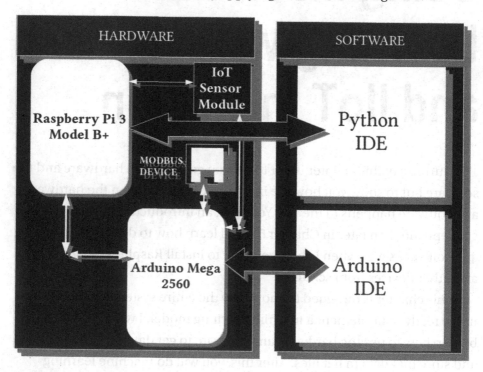

Figure 3-1. *Block system diagram*

You can clearly see the two blocks in the system diagram named Hardware and Software. It is important to distinguish between them so you get a clear understanding of the way the IoT- and IIoT-based solutions work. The software layer communicates with the respective hardware component with its own programming integrated development environment (IDE). The Python IDE here is the Thonny IDE on Raspbian; it's the software component that communicates with the Raspberry Pi 3 Model B+, and the Arduino IDE software component communicates with the Arduino Mega 2560 microcontroller. In the hardware section, this is the most important one to understand. The Raspberry Pi 3 Model B+ is the SBC that acts as a master and hosts the software for Python and Arduino. The software components fully run on the Raspbian. The Raspberry Pi hardware component has GPIO pins inside its board through which it communicates with IoT sensors (remember from Chapter 2 that most IoT sensors are common to Raspberry Pi and Arduino). Raspberry Pi also communicates with the Arduino Mega 2560 via its USB serial port cable and fetches data from it. Please remember that you can add many more devices to the Arduino, especially ones that need Modbus communication to the microcontroller board, than the Raspberry Pi because it has more pins available on its board. The Arduino Mega 2560 communicates with the Modbus devices such as energy meters, etc. Then it gets the data to Raspberry Pi 3 Model B+ back through its USB serial port. In order to create a robust system, you will need to test the communication from the Raspberry Pi Python code and then to its connected IoT sensors or LEDs. After this, you will test the Arduino communication between the Raspberry Pi and Arduino. Once you get the IoT sensor data, you will store it in a lightweight database, SQLite3. In the real world, you could store this in any other database such as Oracle, db2, PostgreSQL, etc. Once you have the data, you need to then apply the machine learning process on top of it to get any meaningful insights into the workings of IoT or any of the connected Modbus devices. In this chapter, however, you will not be using a Modbus device with Arduino but a simple IoT sensor attached to the Arduino microcontroller to test if the serial communication is happening between the master and slave. Let's go ahead and start testing the system.

Testing the System

Before you start your system, you need to ensure that every device in your solution is working properly. Raspberry Pi is your master system and you need to boot it up first to see if it works properly. In order to do so, you need to plug the power cord into the power cord of the Raspberry Pi board.

Once done, you need to connect the LED screen using the HDMI port. All this is done after you have installed the Raspbian OS on the SD card of the Raspberry Pi 3 Model B+. Figure 3-2 shows the Raspberry Pi booting up.

Figure 3-2. *Raspberry Pi bootup*

After it has finished booting up, the Raspberry Pi desktop will show up and will look similar to what you see in Figure 3-3.

Figure 3-3. *Raspberry Pi 3 Model B+ desktop after booting up*

Go to the Start button at the top of your Raspberry Pi desktop, click Programming, and then click the Thonny Python IDE. This is the IDE that you will be using for your programming. Please remember that this is not a very professional IDE but a small one to help you get started on IoT projects. It does not have advanced features that an IDE like PyCharm has, but it is good enough to write code and get it executed on Python IDLE.

Once the Thonny IDE shows up, you'll type a simple "Hello world" program to check if Python's pandas library is installed correctly and working (Listing 3-1). The code for the simple program to test if the pandas library is installed and Python is working properly, hello_world.py, is shown in Figure 3-3. You're importing standard libraries into the program: pandas for managing dataframes, matplotlib for visualizations such as graphs and charts, numpy for mathematical computations, and Seaborn for data visualizations such as heatmaps.

Listing 3-1. hello_world.py

```
import pandas as pd
import matplotlib as plt
import numpy as np
import seaborn as sb

print("Hello World from Raspberry Pi")
```

Run the Python code in the Thonny IDE by pressing the Run button in the toolbar and you should have the output that appears in Figure 3-4.

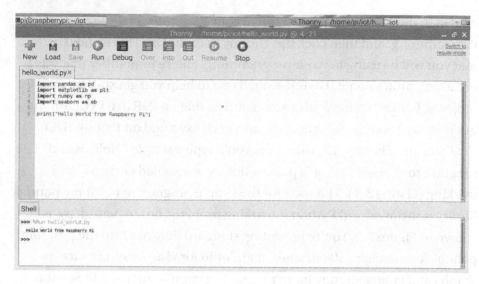

Figure 3-4. *"Hello world" Python code on Raspberry Pi 3 Model B+*

Successful output means that most of the common libraries like pandas, matplotlib, numpy, and Seaborn are installed and that Python IDLE is configured to give you output. If you encounter any error in this simple Python program on the Raspberry Pi, it could be due to a typo or a library not being installed on your Raspbian-based Python. In this case, you could try installing it using the pip install <library_name> command. Replace <library_name> with the name of library that is giving you the error, such as pandas, Seaborn, matplotlib, etc.

You could run another simple test on pandas by creating a test dataframe such as one given in the following code, which imports the pandas dataframe and then creates an instance of the DataFrame() object and then prints out "Pandas loaded". The result appears in the Thonny IDE in Figure 3-5.

```
import pandas as pd

df=pd.DataFrame()
print("Pandas loaded")
```

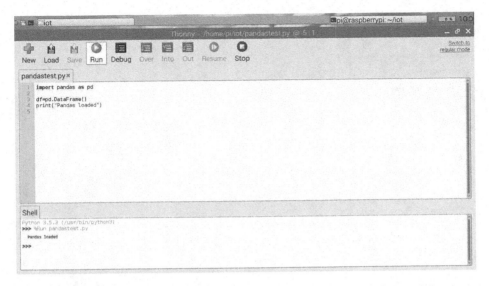

Figure 3-5. *Result of code execution for testing the loading of the pandas dataset*

After running the pandas load test, you are assured that the pandas Python installation is fine. Now you need to test the machine learning library scikit-learn, which is the library you will be using for machine learning implementation in the solution exercises of this book. Run the following code to test scikit-learn loading on Raspberry Pi. The resulting output is shown in Figure 3-6.

```
import sklearn as sk
print("Scikit learn loaded")
```

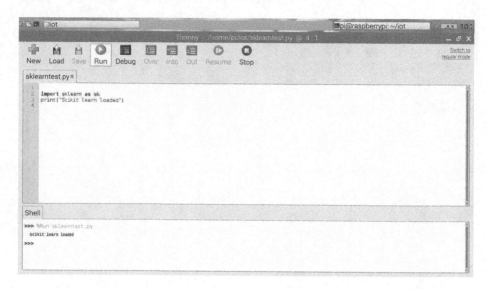

Figure 3-6. *Result of loading scikit-learn on Raspberry Pi*

If the scikit-learn library is loaded properly, you should see the
output given in Figure 3-6 in the shell section at the bottom. If you
get any error from the shell, it means the library is not loaded in
your Raspberry Pi system. To load the library, use the command
`pip3 install scikit-learn`. If you need to troubleshoot, refer to the
Stackoverflow.com discussion on the topic at `https://stackoverflow.`
`com/questions/38865708/how-can-i-run-python-scikit-learn-on-`
`raspberry-pi`.

You now have Python and its libraries tested and running. If everything
has worked fine so far, the next step is to test Arduino using Python, which
you will do in the next section.

Testing Arduino Using Python

Arduino is a microcontroller. As you know from Chapter 2, it is used for
communicating with industrial devices using the Modbus protocol. Let's
get started with Arduino.

Arduino programs are written in the Arduino IDE. The Arduino IDE is special software running on the Raspberry Pi system. You are going to use it as a master, and Arduino is going to be the slave that communicates with your IIoT devices such as energy meters. It allows you to write sketches (a synonym for *program* in the Arduino language) for different Arduino boards. The Arduino programming language is based on a very simple hardware programming language called Processing, which is similar to the C language in its syntax. After the sketch is written in the Arduino IDE, it should be burnt onto the Arduino chipboard for execution. This process is known as uploading to the Arduino microcontroller. In this test, you are going to use the Arduino IDE for testing on a Raspberry Pi 3 Model B+. Please remember Arduino is not a fully functional computer; it does not have its own OS like Raspberry Pi has and that is why it needs a master like Raspberry Pi to monitor, control, and use it. The slave Arduino communicates back and forth between its master and the connected IIoT or IoT devices. Although this looks complicated initially, it becomes easier once you start to test and put together each of the pieces of hardware and software for this system to work.

Arduino Hardware Setup and Communication

You first connect using the USB serial cable that comes with the Arduino Mega 2560 with Raspberry Pi 3 Model B+, as you can see in Figure 3-7. The USB port of the Raspberry Pi is used to do this and the other end is connected to the serial port of the Arduino board.

Figure 3-7. *Connected Raspberry Pi 3 B+ and Arduino Mega 2560*

In the next step, open up the Arduino IDE as shown in Figure 3-8 and write the program shown in Listing 3-2.

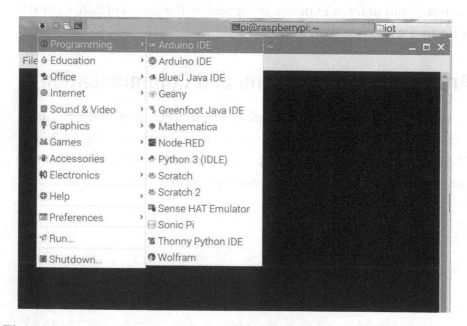

Figure 3-8. *Opening the Arduino IDE on Raspberry Pi 3 Model B+*

Once the Arduino IDE comes up, you can write the Arduino version of the "Hello world" program.

Listing 3-2. Arduino "Hello world" Program

```
void setup() {
  // put your setup code here, to run once:
Serial.begin(9600);
}

void loop() {
  // put your main code here, to run repeatedly:
Serial.println("Hello World from Arduino");
}
```

When you write a sketch program for Arduino, there are two functions that are very similar to the C language that need to be present. The first one is the void setup() function, which is primarily used for initializing things like serial bus communications, LEDs, or other devices connected to the Arduino microcontroller. The Arduino IDE shown in Figure 3-12 is what you will open on Raspberry Pi 3 B+ from the Start ➤ Programming menu from the desktop and then you'll write the program shown in Figure 3-9.

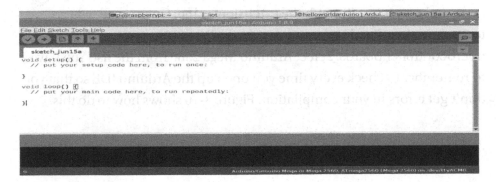

Figure 3-9. *Default Arduino IDE sketch*

The code is self-explanatory and is in the C style and syntax. The first function is `void setup()` and the comment after the symbol `//`, which stands for comment, explains that you can put your code here and that you need to run it only once, like initializing any devices or variable values, etc. In the setup function, the body denoted by curly braces, { }, is the area where you write your code. The next function is the loop function, which you want to run in a loop or repeatedly. You may find this confusing if you have a pure software background and may not understand why you need a function that does things repeatedly. Try to understand it this way: a machine or a microcontroller like Arduino cannot function on its own unless there is a program that tells it what to do again and again. It is like having a servant at your disposal who gets bored of staying idle. Similarly, Arduino also needs to do something repeatedly and cannot stay idle. You can ask it to monitor a device and trigger something when a certain event happens or the device is about to malfunction. This is just an example of what you can do with this `loop()` function; the applications are limitless. You could turn an LED on or off based on a Modbus device's input value such as temperature etc. This just requires that the Arduino microcontroller works continuously without stopping, and that is what this `loop()` function is all about.

Before you can continue, you need to know the process of compiling and running the program in the Arduino IDE. First, you write the sketch code and save it in a file with the extension `*.ino`. Once saved, you go to Tools ➤ Board and hover to open up the list of supported Arduino microcontroller boards. Select Arduino Mega 2560 from the list. You have to remember to check every time you open up the Arduino IDE so that you don't get errors in your compilation. Figure 3-10 shows how to do this.

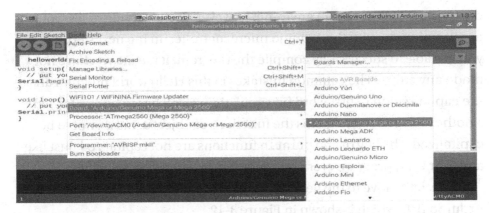

Figure 3-10. *Choosing the Arduino Mega 2560 board*

After choosing Arduino/Genuino Mega or Mega 2560, you are good to write your first Hello world program, as shown in Figure 3-11.

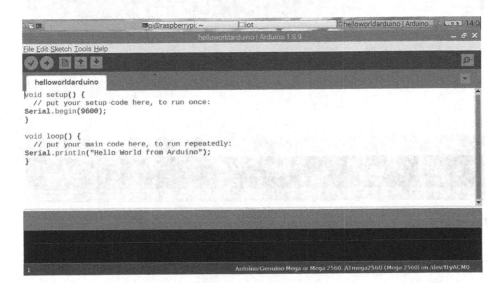

Figure 3-11. *Writing the Hello world program in the Arduino IDE*

You will notice that the setup function of the program initializes the serial port with a baud rate of 9600. The loop function contains a print statement to the serial port of the Arduino. In practical terms, this type

57

of program does not mean anything; however, you are trying to test your communication with the Arduino microcontroller. In the next step, if you are able to successfully compile the program, it means you have not made any errors. The common mistakes in this Hello world program are the capitalization of first word (in serial, the letter S has to be capitalized). Another thing to remember is the functions of object serial do not to be capitalized. The `begin` and `println` functions are not capitalized. Just like any C or C ++, program every sentence has to end with a semicolon.

Now let's move to the next step of compiling the program in the Arduino IDE, which is shown in Figure 3-12.

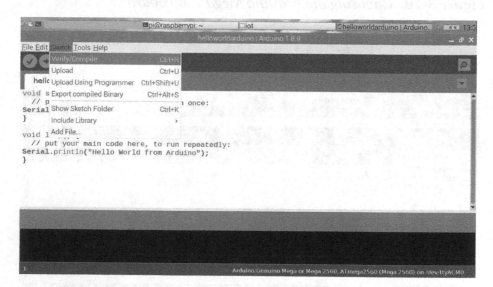

Figure 3-12. *Compiling a sketch in the Arduino IDE*

Running the Sketch

After you have successfully compiled the Hello world program, you may proceed to the next step of uploading it. In this step, the Arduino IDE communicates to the Arduino Mega 2560 microcontroller through Raspberry Pi's USB serial interface and flashes the entire program to it. Now this program will reside in the microcontroller's memory and run infinitely until you write your next program and write it on top via the same process. Figure 3-13 shows the Upload option in the Sketch menu for uploading the program onto the microcontroller board.

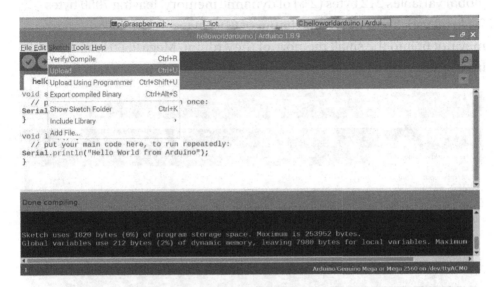

Figure 3-13. *Option in the Arduino IDE to upload the sketch*

Now that you have compiled the program and clicked upload, you should see the program being uploaded and a progress bar which will indicate the progress of the Arduino IDE writing the program to the microcontroller board. If you have not chosen the right microcontroller board, you will see an error here. You can also see an error in this step if there is a problem with the connection between the Raspberry Pi and Arduino. The most common reason for an error at the upload stage is a loose cable or a bad serial bus cable. So check the ends on both Raspberry Pi and Arduino boards to verify your cable does not have any issue. Try changing it if you still face an error when uploading. Figure 3-14 shows the result of a successful upload of your Hello world program onto the Arduino microcontroller board. Notice the "Done Compiling" message at the bottom panel of the Arduino IDE. The box below shows messages related to the program storage space or the maximum space available for it to use. In this case, it has used 1820 bytes of the available maximum space of 253952 bytes. Similarly, it also informs you that it uses for its global variables 212 bytes (2%) of dynamic memory, leaving 7980 bytes for local variables out of a maximum of 8192 bytes. Very long programs may not fit into the small memory of the Arduino Mega 2560 board, so this information comes in handy so you can optimize the program to fit into it.

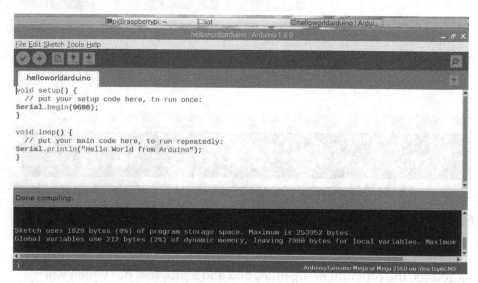

Figure 3-14. *The "done compiling" message on the Arduino IDE*

This is the last step of the process to test the Raspberry Pi to Arduino communication. Now you are ready to undertake the next step in getting IoT sensor data from the Raspberry Pi.

Getting IoT Sensor Data with Raspberry Pi Sample Code

In order to get IoT sensor data in the Raspberry Pi, you first need to relook at the GPIO pins on the SBC microcomputer board. They are shown in Figure 3-15.

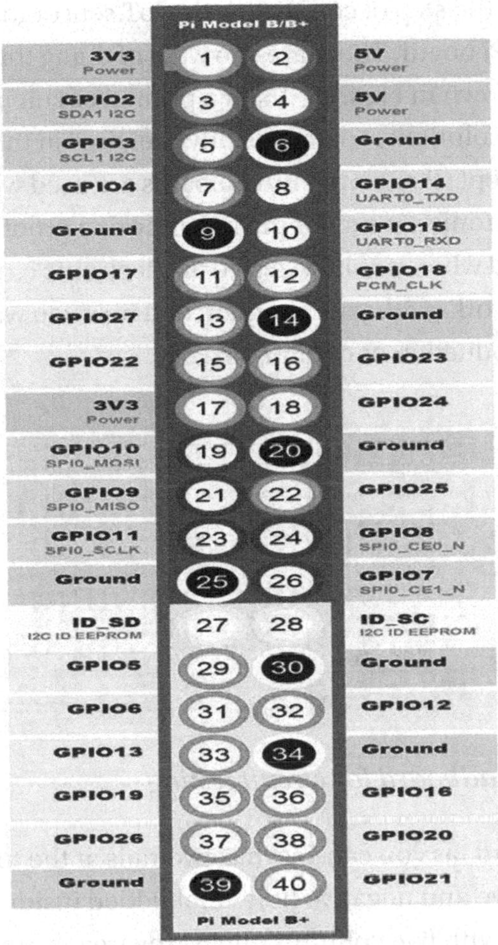

Figure 3-15. *Raspberry Pi 3 Model B+ GPIO pin layout*

The 40 GPIO pins are clearly visible in Figure 3-15. I explained their use in earlier chapters; however, in this chapter you are going to use them to make a circuit to communicate with things like LEDs and IoT sensors. I will refer to these pin numbers henceforth like, for example, the pin with a red circle with number 1 has 3V marked against it, which means it is to be used to power your device with a 3V current and a pin with 5 volts of power has a red circle, like numbers 2 and 4. Also, you will use the ground pins circled with numbers 6, 9, 14, 20, 30, 34, and 39. The rest of the pins can be used for digital GPIO communication.

You're now at the step of connecting the IoT sensor to the GPIO pins of the Raspberry Pi board. There are two ways of doing this: the use of a breadboard, as shown in Figure 3-16, or a printed circuit board (PCB). All the examples, solutions, and case study exercises in this book are at the proof-of-concept (PoC) level. Breadboards are used when you are designing an electronic system for the first time as a proof of concept, and PCBs are used when you have tested an electronic circuit on the breadboard that works with your program and now you want to make it permanent for production or commercial use.

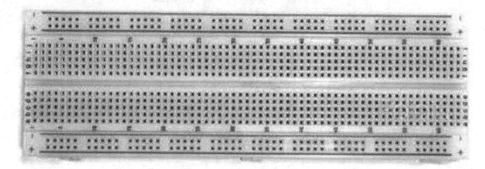

Figure 3-16. *Breadboard for circuit testing*

The breadboard, as you can see, has two rails at the top and bottom which have positive and negative holes embedded inside them. The middle partitions with five columns and 64 rows each are used to connect devices such as LED, IoT sensors, or motors. The key thing to note about this breadboard is that underneath the plastic coating is a mesh of circuitry that connects each of the rows together. So if you place a leg of an LED at any middle portion of the row, you should be able to connect to it by placing another wire on the same row. You need not put the wire on the same hole that you insert into the hole the leg of any device. This feature of the breadboard makes it very convenient to work with and does away with the need to solder the circuit together (which is required when working with printed circuit boards). Since the nature of this board is that the entire circuit is based on holes and it is not permanent, you cannot use it for production-level work.

Connecting It All Together

Let's now connect an IoT sensor on the Raspberry Pi 3 Model B+ board and write a Python program to get data from it. You will learn how to store the IoT sensor data in a database in the next section of this chapter.

The circuit diagram of the IoT sensor with Raspberry Pi needs to be defined first so that you can understand what you are trying to build. Figure 3-17 shows the electrical connections diagram for the project.

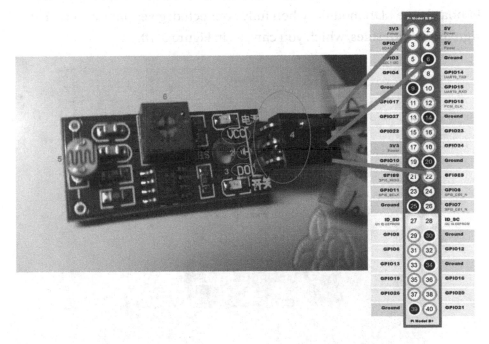

Figure 3-17. *Electrical connections diagram of an IoT sensor connected to the Raspberry Pi*

The IoT sensor that you are going to use is an LDR, or light-dependent resistor module. You're not using an LDR or a photoresistor but a pre-made LDR module that has a completely functional circuit board made on its PCB with the far end hosting the photoelectric resistor marked as number 6 in Figure 3-17. This module senses light from the ambient environment and gives back data in the 0/1 format. The values it gives are in a float, and it gives back 0 for daylight and a value closer to 1 (like 0.90 to 0.99) when it is dark. To simulate darkness, you can use an object to cover the light photo resistor of the LDR. You can see the point marked as 1, which is the VCC; this is where you have to connect the wire to the GPIO pin 1 on the Raspberry Pi board. The point marked is the ground connection and must be connected to the GPIO pin 6 on the Raspberry Pi board. Point number 3 is the DO, or digital output signal wire, from the LDR module and it must be connected to GPIO pin 25 on the Raspberry Pi board. The LDR module, when fully connected, gives out two red LED signals on both sides, which you can see in Figure 3-18.

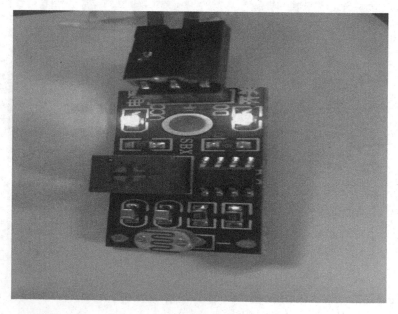

Figure 3-18. *Fully connected LDR module*

Now let's look at the LDR module with the photoelectric resistor. At the far end, it's covered; as a result, only one light glows, showing there is darkness around the LDR module (see Figure 3-19).

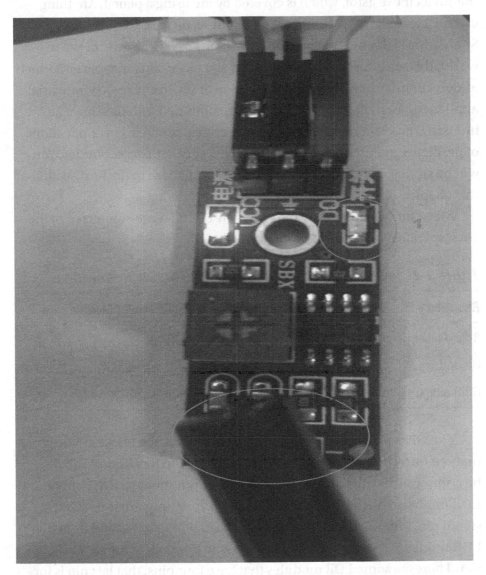

Figure 3-19. *Fully connected LDR module covering the PE resistor*

The point marked 1 in Figure 3-19 is the second LED for digital output, which is off and shows there is no signal being received from the photoelectric resistor to it. The point marked 2 in the diagram is the photoelectric resistor, which is covered by me using a pencil. Anything can be used to cover the PE resistor as long as no light goes through it. Simulation of day and night is for testing purposes; however, in the real world, this sensor works wonders when you place it in an application such as detecting day or night and switching on or off an consumer/industrial appliance based on the result. This data, combined with data from an industrial machine, also helps diagnose critical problems in applications of predictive maintenance, which you will apply in the case studies. You will be storing this LDR IoT sensor data in a database in the next section of this chapter.

Make sure that the GPIO pins listed in Table 3-1 are connected to the corresponding pins on the Raspberry Pi board.

Table 3-1. *Raspberry Pi GPIO Pins to LDR Module Connections*

Raspberry Pi	GPIO Pin Number	Light Sensor Module
3.3v Power	1	VCC (V)
Ground	6	GND (G)
GPIO signal pin	25	SIGNAL (S)

The GPIO pin number 1 has to be connected to the LDR module pin marked on its PCD as VCC. Similarly, the GPIO Pin for Ground should be connected through the wire to the middle pin showing GND on the LDR module. The third pin you are using in your program is the GPIO signal pin, which will send and receive the digital signal to and from the LDR module, which is pin number 25, to the third pin of the GPIO signal pin. There are some LDR modules that have four pins; that last pin is for analog input and output in case you want to use it. But in your case, since you have a three-pin LDR module, you don't have to worry about it. The connected LDR pins on the GPIO and the LDR module can be seen in Figure 3-20.

Figure 3-20. *Raspberry Pi GPIO pins to LDR module connections*

Programming the IoT Sensor LDR Module

The first sign that you have connected the pins correctly is that you will see two red LEDs glowing on the LDR module, signaling that the module is receiving power and input from the Raspberry Pi board. This is essential because unless you get these lights, you cannot implement the Python code shown in Listing 3-3, called ldr.py.

Listing 3-3. Programming the LDR Module IoT Sensor

```
#Turn on Light Sensor
from gpiozero import LightSensor
ldr = LightSensor(25)
while True:
```

```
print(ldr.value)
ldval=float(ldr.value)
#print(ldval)
#print("done deal")
if(ldval>0.0):
        print("It is Night time now")
else:
        print("It is Day time")
```

The library gpiozero is not installed in the default version of Python on Raspberry Pi so you must do so by typing pip install gpiozero in the installed Python directory. This library has the requisite properties and functions to communicate with the GPIO pins and devices. In this program, you first import the gpiozero library LightSensor and then in second line you initialize it to GPIO pin number 25 to get the signal from the LDR module. Refer to Table 3-1 for the GPIO signal pin number to the one given in the second line of this program; they should match or you will not be able to communicate with the LDR module at all. Next, you use an infinite loop to print the LDR value, which is returned as 0 for day and 1 for night. However, the output is a floating point number closer to 1 between 0.9 and 1. If you want it to be more sensitive, you may need to adjust the blue potentiometer on top of the LDR module. This small program uses an if condition to check if the value of the LDR is greater than 0 and then it prints "It is Night time now;" otherwise, it prints "It is Day time." After the loop comes to the end, you turn off the LDR module sensor so that it can be initialized again. In the real world, this is where your application code to control any device such as an LED light or any other appliance through the Raspberry Pi would sit. Listing 3-4 gives the output of the program run in Listing 3-2 using Python.

Listing 3-4. Output of LDR Python Program

```
pi@raspberrypi:~/ python ldr.py
0
0.94490146637
It is Night time now
0.94490146637
It is Night time now
0.94490146637
It is Night time now
0.94490146637
It is Night time now
0.94490146637
It is Night time now
0.94490146637
It is Night time now
0.94490146637
It is Night time now
0.94490146637
It is Night time now
0.94490146637
It is Night time now
0.94490146637
It is Night time now
```

The output is a trial run of the LDR Python program and it works by giving out the floating point values closer to 1, which means it is dark when this program is being run or there is darkness around the LDR sensor module (it may be covered by an object). If you were to light a torch or bulb over the photoelectric resistor of the LDR module, it would start giving a value of 0, indicating there is light around it. You can test this and improve the sensitivity by adjusting the blue square-shaped potentiometer on top of the LDR module PCB. You have come to the end of this section because you are able to achieve getting data from IoT sensors.

Please remember there are many more sensors with many uses and which can be used for simple to complex consumer and industrial applications. The data from any of these sensors would be very similar. The digital output is always on the extreme ends of 0 and 1; however, analog output can be varied and give you in-between readings, which are not in the extremities of 0 and 1. Also, one point to remember is that the accuracy of these IoT sensors deteriorates over time when they are used heavily so you may need to replace sensors that come in contact with water or soil, leading to corrosion of their sensor heads.

Storing IoT Sensor Data in a Database

In this section, I am going to show you how to store data in the SQLite3 database. You will look at how to install this database in Chapter 5 in detail. However, as with previous examples in this chapter, I am making you sit back and have a look at what you can do with the IoT sensors by testing them and putting the system together one by one and accumulating data in a database. In the example in the previous section, you were able to get the LDR module IoT sensor data successfully; however, such data if collected in isolation has no meaning and needs other sets of data to become meaningful. As an example, just knowing whether it is day or night is not enough; if you couple it with temperature IoT sensor data, you can put together a correlation between them to see if light has any relationship to temperature. It would then have some meaning if there was any relationship found.

You will also modify and add new code to the existing code from Listing 3-2 to store it in a SQLite3 database. You can store this data in any other RDBMS or in a flat file like a CSV or JSON format as well; however, here you will use SQLite3 which works well on raspbian.

Configuring a SQLite3 Database

In Figure 3-21, you can see the SQLite3 database named iotsensor.db. Since this is a vanilla database, the SQLite3 database engine automatically creates it for you. This new database will be empty and will not have any tables or other structures. You will need to create them.

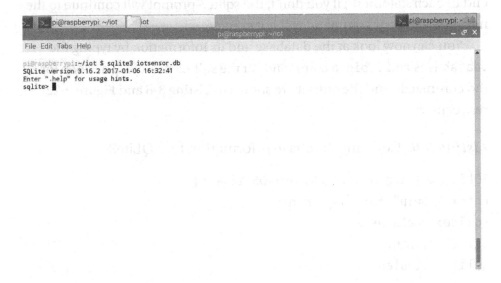

Figure 3-21. *Starting the SQLite3 database*

The code for starting the SQLite3 database is shown in Listing 3-5.

Listing 3-5. Starting the SQLite3 Database

```
pi@raspberrypi:~/iot $ SQLite3 iotsensor.db
SQLite version 3.16.2 2017-01-06 16:32:41
Enter ".help" for usage hints.
sqlite>
```

Once you are logged into the database, it greets you with its version number and the date and timestamp when you have logged in. It displays a prompt sqlite>, which is where you run your commands to work with the databases. There are two sets of commands that work in this prompt: one is the set of commands that start with a dot, such as .databases or .tables, and the other is the set of SQL commands such as select * from <tablename>;. A common beginner mistake is to forget to put a dot before the databases or tables commands and then the SQLite prompt throws out an error. For the SQL commands, do not forget to use a semicolon (;) at the end of each statement; if you don't, the sqlite> prompt will continue to the next line and your command will not execute.

You can now look at the database and its information by typing the .databases and .dbinfo commands at the sqlite> prompt. The code for the commands and the output are shown in Listing 3-6 and Figure 3-22, respectively.

Listing 3-6. Code for Database Information for SQLite3

```
SQLite version 3.16.2 2017-01-06 16:32:41
Enter ".help" for usage hints.
sqlite> .databases
sqlite> .dbinfo
sqlite> .tables
sqlite>
```

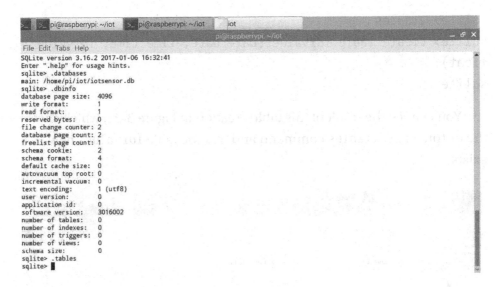

Figure 3-22. *Output of the code for database info commands*

Creating the Database Structure

The output shows various results for the database, such as page size used in memory, text encoding, number of tables, number of indexes, number triggers, and number of views. You do not have any tables, views, triggers, or indexes since this is a vanilla database. You can verify this with the dot command .tables, which will show any tables if they exist, as shown in Figure 3-22. It returned nothing, which means there are no tables in this database. So let's create a table to store your small IoT sensor data with a date and time stamp, as shown in Listing 3-7.

Listing 3-7. Code for Table Creation to Store LDR IoT Sensor Values

```
pi@raspberrypi:~/iot $ SQLite3 iotsensor.db
SQLite version 3.16.2 2017-01-06 16:32:41
Enter ".help" for usage hints.
```

```
sqlite> .tables
sqlite> create table ldrvalues(date date, time time, ldrvalue
float);
sqlite>
```

You can see the result of this table creation in Figure 3-23, which shows the output of the .tables command and that the table ldrvalues now exists.

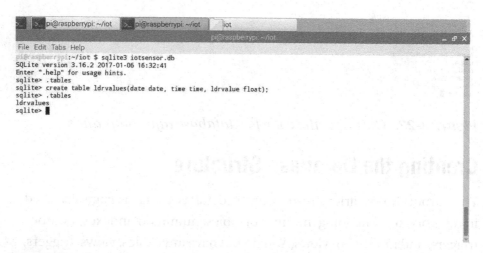

Figure 3-23. *Table creation in SQLite3 for storing IoT sensor data*

Inserting Data into the Database

You now have a table structure named ldrvalues to store your IoT sensor data in a SQLite3 database named iotsensor.db. You can go ahead and update the Python program from Listing 3-2, which fetches data from the IoT sensor, to now write the values into this SQLite3 database. The program is shown in Listing 3-8.

Listing 3-8. Program to Store LDR Module IoT Sensor Data in a
SQLite3 Database

```
pi@raspberrypi:~/iot $ cat ldrdb.py
#Light Dependant Resistor Module Initialization
from gpiozero import LightSensor

#importing Sqlite3 python library to connect with database
import SQLite3

from datetime import datetime

#GPIO LDR Signal Pin initialization
ldr = LightSensor(25)

#Read value infinitely in a loop
while True:
        print(ldr.value)
        ldval=float(ldr.value)
        if(ldval>0.0):
                print("It is Night time now")
        else:
                print("It is Day time")

        conn = SQLite3.connect('iotsensor.db')
        curr=conn.cursor()
        query="INSERT INTO ldrvalues(date,time,ldrvalue)
        VALUES(" +"'" + str(datetime.date(datetime.now()))
        + "'" +"," + "'" +                 str(datetime.
        time(datetime.now())) + "'" + "," + "'" + str(ldval) +
        "'" + ")"
        print(query)
        curr.execute(query)
        conn.commit()
```

Notice the modifications to the code from Listing 3-2. You added import statements to the SQLite3 Python library; this is necessary to communicate with the SQLite3 database. You also imported the datetime library to get date and time so that you can enter it in the database when inserting the query. The while(true): loop didn't change; you just added the database insertion code at the end of it. The conn object is used to connect to the iotsensor.db database, where you created a table named ldrvalues; refer to Figure 3-23 for this. The cursor object inside the while loop initializes the cursor with the connection to the iotsensor database. The next statement is an INSERT INTO... statement which enters data into all the three columns: date, time, and ldrvalue. The value entered into time column is taken from the datetime.now() function, which returns the date and time together. Since you want the time and date separately, you first use datetime.date() and then in the second column, you use the datetime.time() function to get only the time value out of it. You separate the date and time instead of creating a single column because when you are doing EDA it becomes convenient if your date and time are in separate columns; it's easier to find trends with date and time. Although you can do the same operation for taking out date and time from the column during EDA, you save yourself some cumbersome functions at the time of querying it. In a practical world, however, you may rarely find these two values of date and time separate and in the same column known as a timestamp. After the query is done, you use the curr.execute(query) to get the insert into the statement executed by the SQLite database engine. This is the stage when you are likely to get an error if you have a syntax problem in your query.

Checking the Data for Sanity

Based on the error messages that you get, you should try to resolve it. I ran the program for a few minutes and it inserted the LDR module IoT sensor data into the SQLite3 iotsensor.db database. You can see the result in Figure 3-24.

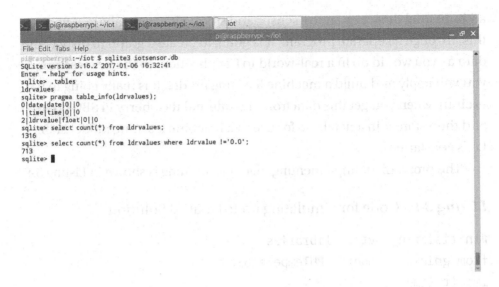

Figure 3-24. *Program execution result of ldrdb.py*

You can see that the program has inserted about 1316 rows in a few minutes. The structure of the table can be seen in the execution of the pragma table_info(ldrvalues); statement. The pragma statement has a function named table_info and takes the argument of the table name of the connected database. In your case, the connected database is iotsensor.db and the table is ldrvalues. It has three columns, which you can see as an output of the pragma statement. The first one is date, the second is time, and the third is ldrvalue. Their respective datatypes in the SQLite3 table are given beside each of them. After this, you run a query of select count(*) from ldrvalues to get the count of total rows. Next, you look at the count of all the values in ldrvalue column which are not equal to 0.0, which stands for day. Any value in this column shows darkness or night. You should understand through the demonstration of this simple program how the IoT sensor data can be stored in databases and used.

Next, I am going to show you a full-fledged Python program that uses the internal Raspberry Pi CPU to measure its temperature and store it in a database. The program is going to simulate that it is measuring data from an industrial device or an IoT sensor and storing it in a database. After this, through a separate program, it is going to create a graph to display the data as you would do in a real-world IoT application. In the next section, you will apply and build a machine learning model. It is really going to be exciting when you get this data from the internal Raspberry Pi SBC board and then store it in a database for use in a machine learning model later. Let's get started.

The program for implementing machine learning is shown in Listing 3-9.

Listing 3-9. Code for Simulating an IoT-Based Solution

```python
#Initializing python libraries
from gpiozero import CPUTemperature
import time
from datetime import datetime
import pandas as pd
import psutil
import platform
from gpiozero import LED
import time

#Initializing Audio for RED status play
import pygame
pygame.mixer.init()
pygame.mixer.music.load("example.mp3")

#intializing LED at pin numberf 18
redled= LED(18)
greenled= LED(22)
yellowled= LED(17)

#Initializing Light Sensitive Module
from gpiozero import LightSensor
ldr = LightSensor(25)
```

```
tempstatus=""

#Columns for pandas dataframe
columns=['date','time','temperature','tempstatus','cpupercent',
'diskpercent','memorypercent']
#Creating a pandas dataframe to store values from Raspberry Pi
hardware
df=pd.DataFrame(columns=columns)
df['date']=datetime.date(datetime.now())
df['time']=datetime.time(datetime.now())
df['temperature']=0
df['tempstatus']=""
df["ldrval"]=0
cpu = CPUTemperature()
counter=0
while True:
        #print(cpu.temperature)
        time.sleep(1)
        tem=cpu.temperature
        if(tem>60):
                print("RED ALERT CPU EXCEEDING HIGH
                TEMPERATURE")
                tempstatus="RED"
                redled.on()
                greenled.off()
                yellowled.off()
                pygame.mixer.music.play()

        elif(tem>55 and tem<60):
                print("YELLOW ALERT CPU NEARING HIGH
                TEMPERATURE THRESHOLD")
                tempstatus="ORANGE"
```

```
            redled.off()
            greenled.off()
            yellowled.on()
    else:
            print("TEMPERATURE IS NORMAL")
            tempstatus="GREEN"
            #time.sleep(1)
            greenled.on()
            redled.off()
            yellowled.off()
    df['date'].loc[counter]=datetime.date(datetime.now())
    print(datetime.date(datetime.now()))
    df['time'].loc[counter]=datetime.time(datetime.now())
    df['temperature'].loc[counter]=tem
    df['tempstatus'].loc[counter]=tempstatus
    #print(df['date'].values)
    #print(df['time'].values)
    #print(df['temperature'].values)
    #print(df['tempstatus'].values)
    #Now write data in database SQLite3 temperature.db
    #print("Connected to MACHINEMON Database")
    import SQLite3
    conn = SQLite3.connect('machinemon.db')
    #df.to_sql(name='tempdata', con=conn)
    curr=conn.cursor()
    #get machine data

    os, name, version, _, _, _ = platform.uname()
    version = version.split('-')[0]
    cores = psutil.cpu_count()
    cpu_percent = psutil.cpu_percent()
```

```
memory_percent = psutil.virtual_memory()[2]
disk_percent = psutil.disk_usage('/')[3]
#Getting Light Sensor Data to determine day or night
values 0 means Day and 1 means Night
#print(ldr.value)
ldrval=ldr.value
#boot_time = datetime.datetime.fromtimestamp(psutil.
boot_time())
#running_since = boot_time.strftime("%A %d. %B %Y")
#query="INSERT INTO TEMPERATURE VALUES(" +"'" +
str(datetime.date(datetime.now())) + "'" +"," + "'" +
str(datetime.time(datetime.now())) + "'"+ "," + "'" +
str(tem) +  "'" + "," + "'" + tempstatus +"'" + ")"
query="INSERT INTO machinedata(date,time,temperatur
e,tempstatus,cpupercent,diskpercent,memorypercent,l
drval) VALUES(" +"'" + str(datetime.date(datetime.
now())) + "'" +"," + "'" + str(datetime.time(datetime.
now())) + "'"+ "," + "'" + str(tem) +  "'" + "," + "'"
+ tempstatus + "'" + "," + "'" + str(cpu_percent) + "'"
+ "," + "'" + str(disk_percent) + "'" + "," + "'" +
str(memory_percent) + "'" "," + "'" + str(ldrval) + "'"
+ ")"
print(query)
curr.execute(query)
conn.commit()
#Increment counter to parse to next record number in
the dataframee
counter=counter+1
```

Although what you see is a large chunk of code, the real purpose is
to show you how a practical Python-based IoT application is built. The
only simulation you are doing is instead of getting data from an actual IoT

sensor, you are taking it from the internal hardware IoT sensors embedded in the Raspberry Pi. The concept of temperature and the percent values are applicable to industrial devices such as heat exchangers or boilers, which have a mix of values in temperature and percentages and some absolute values as well. The electrical devices have various values ranging from frequencies to voltages. All this makes up the data you will use in the case study examples because there the data is going to come from the actual IoT sensors connected to the Raspberry Pi and Arduino.

The code first initializes and imports the required Python libraries like CPUTemperature from the gpiozero library. This is the one that is going to get you the internal Raspberry Pi temperature. You also import libraries like datetime to get the time and date of when the data is generated, the pandas dataframe to store the data temporarily, and an imported LED to give a status of green, orange, or red through the lighting of the appropriate one. For example, if the CPU temperature is less than 55 degrees Celsius, the green LED will light up; if the CPU temperature of the Raspberry Pi is greater than 55 but less than 60 degrees Celsius, the orange LED will light up; and if the temperature is greater than 60 degrees, the red LED will light up. This is exactly how you would implement an advance warning system using IoT sensors if you were measuring outside temperature values. After this, you import pygame to ensure that there is a sound coming out of the Raspberry Pi speaker when the temperature reaches beyond 60 degrees Celsius. This is a critical system for your SBC and if something is not done, the system will hang after a while and the board or its components can also burn. Try to imagine this in an IoT-based environment where you would like to implement such a critical alert system. The library pygame is initialized using the init() function and in the next line, it loads example. mp3, a shrill alerting sound, into memory. It does not play it since you have not given the command yet; it is just loaded into the memory.

In the next section of the code, you initialize the LEDs that will light up according to the CPU temperature. The red LED is at GPIO pin number 18, the green LED is at GPIO pin number 22, and the yellow LED is at GPIO

number 17. Make sure the LEDs on the Raspberry Pi have been connected
to the right pin numbers given in Figure 3-15.

The advance warning system now comprises a sound for the red
temperature and a visual alert using LEDs for the CPU temperature. In a
real-world application, you will need these alerting systems plus others
like SMS or e-mail alerts for which you may need to configure mail and
SMS gateway servers on your Raspberry Pi. We will not be doing this as we
are undertaking PoC-level code and this is beyond the scope of the book.

Now you need to initialize the LDR, or light sensor module, at pin
number 25, which you did in an earlier section of this chapter. After this,
you initialize the pandas dataframe to store data in memory. The reason
you are using pandas is that it gives a structure to the data and allows
to you manipulate and analyze it as per need in Python. Next is the
infinite while loop in which first you get the CPU temperature using cpu.
temperature and store it in a variable. The next step is to check the value
of the temperature. If it is greater than 60, an alert is displayed on the
screen saying "RED ALERT CPU EXCEEDING HIGH TEMPERATURE"
followed by the red LED being turned on in the program using the
redled.on() function. When this event happens, the other LEDs (green
and yellow) should be turned off so you use the off() function for these
LED objects so that there is no confusion as to the status of the CPU
temperature. You do not want all of the LEDs to glow; only the red LED
should glow at this time to show the critical status. The pygame is used to
play an alerting sound from the Raspberry Pi speaker using the pygame.
mixer.music.play() function. Similarly, there are two other conditions.
The next is when the temperature value of the CPU lies between 55 and
60 degrees; the status is displayed on the screen as "YELLOW ALERT
CPU NEARING HIGH-TEMPERATURE THRESHOLD" and the yellow
LED is turned on using the on() function and the red and green LEDs
are turned off using their respective off() functions. There is no sound
played because this is not a critical situation; the sound plays only when
the temperature reaches a critical limit of 60 degrees. You may wonder

how I came up with this number of 60 degrees Celsius. I simply referred
to the Raspberry Pi manual, which came with it, and it mentions a range
of 30 to 60 degrees Celsius. The next is the green status where the CPU
is safe and does not need any alerting, but for someone watching the
screen, a message of "TEMPERATURE IS NORMAL" is displayed. The
green LED is turned on using its on() function and the yellow and red
LEDs are turned off using their respective off() functions. Once you
have this status of the CPU system in the variable tempstatus, you need
to store it in a pandas dataframe and this is done by storing the date
and time using the datetime now() function in the df['date'] and
df['time'] columns. df['temperature'] stores the temperature from
the tem variable. The df['tempstatus'] column stores the value that you
get after going through the if condition for the tempstatus variable of
"GREEN," "RED," or "YELLOW."

Now you need to prepare to write the data into the SQLite3 database,
which is done in the import statement import SQLite3. Then you initialize
the connection object to the machinemon.db database. A curr cursor
object is created after this to help parse the table if needed. Before you
can write to the dataframe, it has empty values for some of the columns
like cpu_percent, memory_percent, disk_percent, and LDR value. The
cpu_percent variable is used for storing the CPU percentage value, the
memory_percent variable is used for storing the memory percentage value,
disk_percent is used for storing the disk percentage value, and the ldrval
variable is used for storing the LDR module value of day or night.

Next, you have all the data to write to the database table so now you
construct an insert into statement similar to what you did in the code
in Figure 3-33 using the datetime now() function for date and time,
and appending the values of the rest of the columns like temperature,
cpu_percent, disk_percent, tempstatus, memory_percent, and ldrval
variables, which you got from Raspberry Pi earlier using the psutil library.
Using the curr cursor object, which you created out of the SQLite3 library,
the query string containing the insert into statement is executed using

the execute() function. To count the number of records that have been inserted into the machinedata table, you use a counter variable that is incremented at the end of the while loop as well. Listing 3-10 shows the result of running the code from Listing 3-9.

Listing 3-10. Output of Running machinemon.py

```
pi@raspberrypi:~/iot $ python machinemon.py
TEMPERATURE IS NORMAL
2019-06-19
INSERT INTO machinedata(date,time,temperature,temp
status,cpupercent,diskpercent,memorypercent,ldrval) VALUES
('2019-06-19','22:09:22.857943','47.236','GREEN
','8.1','43.9','29.1','0.0')
TEMPERATURE IS NORMAL
2019-06-19
INSERT INTO machinedata(date,time,temperature,temp
status,cpupercent,diskpercent,memorypercent,ldrval)
VALUES('2019-06-19','22:09:23.914717','47.236','GREEN
','8.2','43.9','29.2','0.0')
TEMPERATURE IS NORMAL
2019-06-19
INSERT INTO machinedata(date,time,temperature,temp
status,cpupercent,diskpercent,memorypercent,ldrval)
VALUES('2019-06-19','22:09:24.970186','47.236','GREEN
','10.2','43.9','29.2','0.0')
TEMPERATURE IS NORMAL
2019-06-19
INSERT INTO machinedata(date,time,temperature,temp
status,cpupercent,diskpercent,memorypercent,ldrval)
VALUES('2019-06-19','22:09:26.038476','46.16','GREEN
','12.3','43.9','29.5','0.0')
```

```
TEMPERATURE IS NORMAL
2019-06-19
INSERT INTO machinedata(date,time,temperature,temp
status,cpupercent,diskpercent,memorypercent,ldrval)
VALUES('2019-06-19','22:09:27.102077','47.236','GRE
EN','13.0','43.9','29.2','0.0')
TEMPERATURE IS NORMAL
2019-06-19
INSERT INTO machinedata(date,time,temperature,temp
status,cpupercent,diskpercent,memorypercent,ldrval)
VALUES('2019-06-19','22:09:28.153815','47.236','GREEN
','12.6','43.9','29.2','0.0')
TEMPERATURE IS NORMAL
2019-06-19
INSERT INTO machinedata(date,time,temperature,temp
status,cpupercent,diskpercent,memorypercent,ldrval)
VALUES('2019-06-19','22:09:29.205080','47.236','GREEN
','14.1','43.9','29.2','0.0')
TEMPERATURE IS NORMAL
2019-06-19
INSERT INTO machinedata(date,time,temperature,temp
status,cpupercent,diskpercent,memorypercent,ldrval)
VALUES('2019-06-19','22:09:30.265720','47.236','GREEN
','9.9','43.9','29.3','0.0')
```

As you can see, this output gives you the insert into statements that happen very fast, almost every second. Data is inserted from the Raspberry Pi hardware sensors and the OS, and then it's stored into the machinemon. db database in SQLite3. Listing 3-11 shows the database structure for storing the internal Raspberry Pi sensor data in the SQLite3 database named machinemon.db and a table named machinedata.

Listing 3-11. Database Structure for Storing Data from machinemon.py

```
pi@raspberrypi:~/iot $ SQLite3 machinemon.db
SQLite version 3.16.2 2017-01-06 16:32:41
Enter ".help" for usage hints.
sqlite> .databases
main: /home/pi/iot/machinemon.db
sqlite> .tables
machinedata
sqlite> .schema
CREATE TABLE [machinedata] ([date] date , [time] time,
[temperature] numeric , [tempstatus] nvarchar(7) ,[cpupercent]
numeric,[diskpercent] numeric,[memorypercent] numeric, outage
varchar, ldrval boolean);
sqlite> pragma table_info(machinedata);
0|date|date|0||0
1|time|time|0||0
2|temperature|numeric|0||0
3|tempstatus|nvarchar(7)|0||0
4|cpupercent|numeric|0||0
5|diskpercent|numeric|0||0
6|memorypercent|numeric|0||0
7|outage|varchar|0||0
8|ldrval|boolean|0||0
sqlite> select count(*) from machinedata;
3606
sqlite>
```

This database structure needs to be created before you can start the execution of machinemon.py because if the back-end database structure does not exist, it will throw an error and the program will fail. Here you first create a new database connection to the database named machinemon.db and connect to it with the statement SQLite3 machinemon.db. Once this is done, in the code you look at basic data by running the dot commands .databases, .tables, and .schema, which tell you about the database's path, the tables that exist within the database, and the schema of the tables or the structure of the tables. In your case, the table is machinedata, which comprises of the following columns: date, time, temperature, tempstatus, cpupercent, diskpercent, memorypercent, outage, and ldrvalue. The uses for these columns were shown in the code execution in Listings 3-9 and 3-10.

Creating the IoT GUI-Based Monitoring Agent

For any professional IoT application to really work as an early warning system it needs to have a monitoring agent in place to monitor the status and send out more alerts such as on a GUI, on the screen, and an email to the administrator. You are going to do just that: create a small monitoring agent for your CPU monitoring program. There are many approaches to creating a perfect monitoring agent; however, to keep this brief, I have taken a simplistic approach using the crontab scheduler on the Raspberry PI, which runs every minute and checks the status of the LEDs and then determines the action to take based on which LED is on at any given point in time. So if the red LED is on, the code will send out a critical alert on the GUI screen and an email to an administrator as well. You can see the code in Listing 3-12.

Listing 3-12. Code for Machinemon Application Monitoring Agent

```
**THIS IS A PYTHON VERSION 3.6 COMPATIBLE CODE
import tkinter as tk
import tkinter.font
from gpiozero import LED
import smtplib
```

```python
win= tk.Tk()
win.title("LED Monitoring Agent Application")
myfont= tkinter.font.Font(family= 'Helvetica', size=30,
weight='bold')
redled=LED(18)
yellowled=LED(22)
greenled=LED(17)

def ledstatus():
    while(1):
        if(redled.value==1):
            print("RED ON")
            sender = 'newsletter@machinelearningcasestudies.com'
            receivers = ['newsletter@machinelearning
            casestudies.com']

            message = """From: From Person
             <mmagent@machinemon.py>
            To: To Person <administrator@machinemon.py>
            Subject: SMTP e-mail CRITICAL ALERT

        Message:      This is a Critical Message alert the CPU
                      Temperature of Raspberry
                      Pi has crossed Threshold value.
            """

            try:
                smtpObj = smtplib.SMTP('mail.
                machinelearningcasestudies.com')
                smtpObj.sendmail(sender, receivers,
                message)
                print("Successfully sent email")
            except smtplib.SMTPException:
```

```
                                    print("Error: unable to send email")
                                            if(yellowled.value==1):
                                    print("YELLOW ON")
                                                if(greenled.value==1):
                                    print("GREEN ON")

def exitprog():
        win.quit()

#command=ledstatus
statusButton= tk.Button(win, text=",command=ledstatus,font=myfo
nt ,bg='green', height=1, width=24)
statusButton.grid(row=0,sticky=tk.NSEW)
exitButton= tk.Button(win, text='EXIT', font=myfont ,
command=exitprog,bg='green', height=1, width=24)
exitButton.grid(row=30,sticky=tk.NSEW)

tk.mainloop()
```

This agent runs on Python 3.6 as the library used is the tkinter
GUI library. To run it, you simply type on the Raspbian command
line python3 mmagent.py. Please note the use of python3 and not just
python. Just typing python will invoke the Python 2.7 compilers whereas
typing python3 will invoke the Python 3.x compiler. To make the code in
Figure 3-38 compatible with Python 2.7, you may first have to install the
Tkinter GUI python library and then change the code to import Tkinter
instead of the small case tkinter. The difference between the uppercase
and lowercase can make it confusing for people who do not understand
the significance. A StackOverflow discussion on this topic may help you
understand this better; go to https://raspberrypi.stackexchange.
com/questions/53899/tkinter-on-raspberry-pi-raspbian-jessie-
python-3-4.

The code in Listing 3-12 creates a window of the instance of the tk.TK() win object. It gives it a title of LED Monitoring Agent Application and then in the next line sets the font for the screen to Helvetica with a size of 30 and a weight parameter of bold. After this, you initialize the three LEDs at GPIO pins on the Raspberry Pi board at pin numbers 18-Red, 22-Yellow, and 17-Green. Next, you create a function to check the LED status because this needs to be done repeatedly each time the agent calls it. The function checks the LED status value through its value property and checks if it is equal to 1. If the value of any of the LEDs is equal to 1, this means the LED is on. It prints the LED status on the screen. After this function, you have another function to help exit the program gracefully. This function has a single line which uses the win.quit() function to exit the application window, bringing down the application monitoring GUI with it. In the general section after this exitprog() function, you instantiate an ledstatus button using the tk.Button() method, which takes in arguments for command=ledstatus. The ledstatus is the name of the button function that is required to be executed when the status button is pressed. By default, the color of the button is green. Then it is set in the grid, telling it using the grid() function for the row where it has to be placed; here you use 0 because you want it to be placed at top of the screen. The sticky argument takes the argument tk.NSEW, which stands for North, South, East, and West. In a similar manner, you now fix the exit button, which allows the user to exit gracefully from the program using win.quit() in the exitprog() function. The last statement of this program is the tk.mainloop(), which is required to make the window stay on the screen until the application work is done. The resultant screen window on the Raspberry Pi after running the command from the command line $python mmagent.py is shown in Figure 3-25.

Figure 3-25. *Running the LED Monitoring Agent Application*

You can run this agent monitoring service application through crontab as well by placing a simple crontab entry, as shown here:

```
$crontab -e
0 10 * * 1 python3 /home/pi/iot/mmagent.py
```

The entry in the crontab above runs the command python3 /home/pi/iot/mmagent.py each day at 10 a.m. and the LED monitoring agent checks the LED status and outputs the values on the screen and sends an alert through email if the status at that time is RED. You can modify the crontab entry to check every few minutes.

In the last section of this chapter, you'll apply a machine learning model on the simulated sensor data that you gathered in your SQLite3 database.

Applying Machine Learning Model on the Sensor Data

To undertake machine learning and apply it on the simulated sensor data from the previous section in the machinemon.db database and the machinedata table, you need to execute the code shown in Listing 3-13.

Listing 3-13. Machine Learning Application on Simulated Sensor Data

```
import pandas as pd
import SQLite3
conn = SQLite3.connect('machinemon.db')
#df.to_sql(name='tempdata', con=conn)
curr=conn.cursor()
#query="INSERT INTO TEMPERATURE VALUES(" +"'" +
date(datetime.now())) + "'" +"," + "'" + str(datetime.
time(datetime.now())) + "'"+ "," + "'" + str(tem) +  "'" + ","
+ "'" + tempstatus +"'" + ")"
df = pd.read_sql_query("select * from machinedata;", conn)
print(df)
#curr.execute(query)
conn.commit()

#Looking at data
print(df.columns)
print(df.shape)
#Looking at datatypes
print(df.dtypes)
df['outage']=df['outage'].astype("int")
#Cleaning up and dummy variables
```

```python
df['tempstatus'] = df['tempstatus'].map({'RED':2, 'ORANGE':1,
'GREEN':0})
df=df.drop('date',1)
df=df.drop('time',1)

#Checking for missing values
print(df.isnull().any())

#EDA- Exploratory Data Analysis
print("----------EDA STATISTICS---------------")
print(df.describe())
print("----------Correlation--------------")
print(df.corr())

#Dividing data into features and target
target=df['outage']
features=df.drop('outage',1)

#Building the Model
from sklearn.model_selection import train_test_split
x_train, x_test, y_train, y_test = train_test_split( features,
target, test_size=0.25, random_state=0)

from sklearn.linear_model import LogisticRegression

lr  = LogisticRegression()
lr.fit(x_train, y_train)
# Returns a NumPy Array
# Predict for One Observation (image)
lr.predict(x_test)

predictions = lr.predict(x_test)

# Use score method to get accuracy of model
score = lr.score(x_test, y_test)
print(score)
```

```
import matplotlib.pyplot as plt
import seaborn as sns
from sklearn import metrics
import numpy as np

cm = metrics.confusion_matrix(y_test, predictions)
print(cm)

plt.figure(figsize=(9,9))
sns.heatmap(cm, annot=True, fmt=".3f", linewidths=.5, square =
True, cmap = 'Blues_r');
plt.ylabel('Actual label');
plt.xlabel('Predicted label');
all_sample_title = 'Accuracy Score: {0}'.format(score)
plt.title(all_sample_title, size = 15);

plt.figure(figsize=(9,9))
plt.imshow(cm, interpolation='nearest', cmap='Pastel1')
plt.title('Confusion matrix', size = 15)
plt.colorbar()
tick_marks = np.arange(10)
plt.xticks(tick_marks, ["0", "1", "2", "3", "4", "5", "6", "7",
"8", "9"], rotation=45, size = 10)
plt.yticks(tick_marks, ["0", "1", "2", "3", "4", "5", "6", "7",
"8", "9"], size = 10)
plt.tight_layout()
plt.ylabel('Actual label', size = 15)
plt.xlabel('Predicted label', size = 15)
width, height = cm.shape
for x in xrange(width):
 for y in xrange(height):
  plt.annotate(str(cm[x][y]), xy=(y, x),
  horizontalalignment='center',
```

```
verticalalignment='center')
plt.show()
```

As you can see from the code, it starts by importing the Python libraries you are going to use in your program later, like pandas and SQLite3. You then initialize the connection to the `machinemon.db` database and create a cursor for the connection object. This time, instead of using a query string to create a query, you use the pandas object `pd` and the function under it, `read_sql_query()`, by passing the sql query and the `conn` object to connect to your database through the SQLite3 driver. The entire `machinedata` table from the `machinemon.db` database is transferred to the pandas dataframe `df` using the code `df = pd.read_sql_query("select * from machinedata;", conn)`.

The next part is EDA where you look at the size and shape of the dataframe `df` using `df.columns` and `df.size` statements. In the next part of the code, you use the `astype()` function to convert the outage to an integer because it contains values of 1 for Yes to outage and 0 for No to outage. The `tempstatus` column contains values that are non-numeric (RED, GREEN, and YELLOW); they need to be mapped to 0 for green, 1 for orange or yellow, and 2 for red. You need to do this because the machine learning libraries in Python can't handle non-numeric values and need numeric data to compute. This conversion is also known as creation of dummy variables through the statement `idf['tempstatus'] = df['tempstatus'].map({'RED':2, 'ORANGE':1, 'GREEN':0})` in the code.

The next part of the code is to check if there are any missing values, which is not the case since you have your own program inputting values into the database and there is no human intervention during this process.

You do exploratory data analysis using the `df.describe()` function of the pandas dataframe object `df`. After this is done, you then look at the correlation between the variables to check if any strong relationship exists. After determining this, you move to divide your dataframe into a target and features. The target is what you want to predict. In your case, it is an outage. You want to predict the outage and see when it happens. You drop outage from the features since it is your predictor.

Now you can start creating your training and testing datasets. For this, you use the `train_test_split` function from the sklearn library. You know that outage, which is your predictor, can have a value of either 1 or 0. 1 means there is an outage and 0 means no outage. So this is a classification problem. You use a simple classification algorithm: logistic regression. You can use others like Naive Bayes or SVM to look at how the accuracy of the prediction improves or not.

You train the logistic regression with `x_train` and `y_train` datasets and then predict using `x_test`. The predictions are stored in the `predictions` variable. To get the accuracy of the model, you use the `score()` function by inputting the `x_test` and `y_test` datasets to it. Once you know the score for the prediction, you can visualize the confusion matrix by using your predictions. This is done with the statement `cm = metrics.confusion_matrix(y_test, predictions)`. The next set of lines initializes the `matplotlib` object and creates a graph figure heatmap by inputting the `cm` (confusion matrix) object into it. You see two graphical visualizations in this section, the first one being the accuracy score and the second one being the confusion matrix that is input into the `plt.title()` function. Lastly, `plt.show()` puts both graphs on the screen. If you run this code in a Raspbian command line window, it will automatically pop up a window with both graphs on the desktop, which will look very much like Figures 3-26 and 3-27.

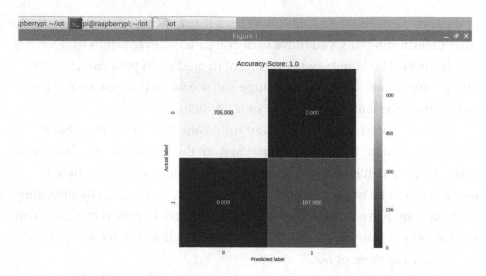

Figure 3-26. *Heatmap of the accuracy score*

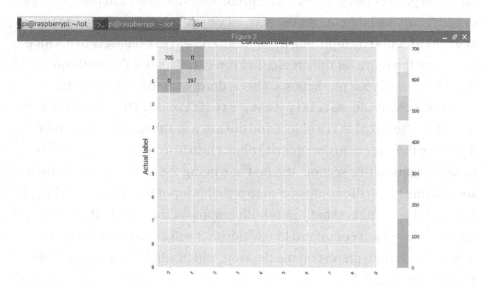

Figure 3-27. *Confusion matrix*

Summary

In this chapter, you looked at Raspberry Pi using Python and the overall block diagram of this IoT monitoring application. You tested your system by writing a Python Raspberry Pi Hello world program. Then you tested to see if all the Python libraries like scikit-learn and pandas were loading perfectly or not. Then you tested your Arduino setup using Python to see if it was communicating properly after setting up its hardware. You ran your first sketch by flashing the Arduino's memory with it. Next, you saw how to get IoT sensor data from Raspberry Pi's sample code by first configuring its GPIO pins and then the hardware setup using the breadboard circuit. You then learned to program the IoT sensor light density resistor module through a Python program to detect day and night. The next important part of the program was to store IoT data in a database like SQLite3 by configuring it and creating a database structure. You then ran a Python program to get the data from the LDR module and store it in the SQLite3 database. Then you created a practical Python-based IoT application with advanced warning capabilities based on data from temperature and light statuses, allowing you to create a monitoring agent for it. Then you applied machine learning on top of this collected data from the temperature and LDR module by using a logistic regression algorithm to classify temperature status based on readings from the IoT sensor. You created a visualization of the classification model by building a confusion matrix using the Seaborn Python library and determining the accuracy of your prediction model.

With the visualization of the accuracy score and confusion matrix, I conclude this section and the chapter. You now have a substantial understanding of how the IoT sensor data can be used to apply machine learning on it to get any business results. In Chapter 5, you will look at the setup and installation required in order to execute IoT and IioT applications using Python.

CHAPTER 4

Using Machine Learning and the IoT in Telecom, Energy, and Agriculture

In Chapter 2, you looked at the IoT and IIoT and their uses. You also saw via a practical example how an IoT system is designed using Raspberry Pi, Arduino, and Python with machine learning. By seeing a practical example, you now have an understanding of the implementation aspects of any IoT or IIoT applications such as creating monitoring systems for giving advanced warning signals to users by gathering data from IoT sensors and then measuring it against a benchmark number such as a temperature threshold and alerting the users through email or GUI alerts.

In this chapter, you are going to see the state-of-the-art implementation of machine learning and IoT in the telecom, energy, and agriculture domains. I will be walking you through some rapid implementations of machine learning and the IoT in each of these sectors, helping you understand the nuances and the complexity involved in building applications for them.

© Puneet Mathur 2020

P. Mathur, *IoT Machine Learning Applications in Telecom, Energy, and Agriculture*,
https://doi.org/10.1007/978-1-4842-5549-0_4

State-of-the-Art Implementation of Machine Learning and the IoT in the Telecom Domain

The telecom domain is going through a rapid technological wave right now all over the world. The heart of this change is the 5G spectrum introduction. In my blog post on two reasons why 5G IoT is safe (`https://pmauthor.com/2019/07/02/3-reasons-why-5g-iot-is-safe/`), I talk about what the spectrum of each telecom technology was able to achieve. Mobile technology has seen four generations of technological development with 1G, 2G, 3G, and 4G. Each generation is a set of telephone network standards for communication across mobile devices such as mobile phones and tablets. The 1G gave 2.4 kbps of connectivity, 2G gave 64 Kbps of connectivity and was based on GSM, 3G gave 144 kbps to 2 Mbps of connectivity, and 4G gave 100 Mbps to 1 Gbps of connectivity and is based on LTE technology. The 1G systems started in Australia and were analog in nature. Truly digital telecommunication started with the introduction of the 2G standard, which brought with it the CDMS and GSM protocols with SMS and MMS services, among others. 3G led to the introduction of smartphone technology such as web browsing, video streaming, and more. 4G has revolutionized the cost and usage of data as we knew it; with 3G, it was very costly. 4G gave high speed, **high quality, and high capacity at affordable rates to users**.

In my blog post at `https://pmauthor.com/2019/07/02/3-reasons-why-5g-iot-is-safe/`, I also pointed out the key benefits of 5G. The biggest benefit that 5G will give mankind, apart from high data connectivity speeds of up to 35.46 Gbps, is the promise of connecting up to 100 billion devices through its network. The communication protocol that is going to enable this is known as mmWave, or millimeter wave mobile communication. This is the biggest breakthrough for the 5G telecommunication standard. We'll look in detail at how this works technically.

Now let's look at the state-of-the-art implementation of machine learning and the IoT in the telecom domain.

HUAWEI: Yes, this is one company in the AI sector to watch out for in spite of the controversy of spying and data breaches that it has been embroiled in. This company was the first one to launch the full mobile 5G chipset. In my blog post on three top machine learning skills for 2019 (`https://pmauthor.com/2019/07/05/3-top-machine-learning-skills-for-2019/`) and another post covering two reasons why 5G IoT is safe (`https://pmauthor.com/2019/07/02/3-reasons-why-5g-iot-is-safe/`), I say that 5G IoT is going to be the next game changer in the world of AI. Huawei has launched a full capability 5G chipset. It has also launched the world's first data center switch with an AI brain. Huawei has also launched full-stack, all-scenario AI technologies to enable an **autonomous driving network** and developed the SoftCOM AI solution to help **mobile operators maximize** energy efficiency, network performance, O&M efficiency, and user experiences. Huawei's solution focuses on building a digital village, as mentioned in the article at `www.busiweek.com/huawei-presents-its-simplified-5g-and-softcom-ai-solutions/`. IoT will also enable its applications to do 8K high-definition (HD) live streaming of audio and video between devices. Real-time virtual healthcare doctors will be able to stream both ways with a patient in critical condition from home, and employees can live stream a business meetings with much clarity and low feedback along with transfer of high speed data. With the launch of 5G services in every country, it's now possible to provide IoT cloud services, personal mobile services, five-star premium home broadband, and cloud-network convergence. Although Huawei does not describe much about its OceanConnect IoT platform on its website (`www.huawei.com/minisite/iot/en/overview.html`), it is not difficult to see the key areas of IoT that it sees developing in society. See Figure 1 on the aforementioned webpage for Huawei's IoT solution.

5G and the IoT are destined to take off together since the technology that the 5G telecom spectrum provides not only compliments the capabilities of IoT but also enhances them by providing very fast and efficient data transfers, which are the backbone of IoT devices and applications in every country where it is implemented.

The OceanConnect IoT platform offers to integrate the IoT at all levels of human life. The base of the OceanConnect IoT platform is the 5G chipset or 5G devices which have their own cloud internetworking gateway on top of 2G for 3G and other IoT and 5G connectivity infrastructure networks, combined with a smart home gateway. On top of this layer is the OceanConnect IoT platform, which has three major components inside it. The first is IoT connection management wherein it sits on top of the 2G, 3G, 4G, or 5G infrastructure that has the smart home IoT gateway sitting in the individual houses of people. The connection management provides access to the OceanConnect IoT platform to these individual smart home gateways. Each of these smart home gateways will have enabled IoT devices and the IoT platform will provide device management services such as adding a device to the IoT platform or removing the device from the IoT platform period even if you have the best infrastructure of a smart gateway combined with connection management and device management services on the IoT platform; however, if you do not have an application enablement capability on the platform, it will not have any practical use. It is the application that actually gives the benefit to the user and provides a stable 5G connection for devices. So this IoT platform provides application enablement services as well. Huawei looks at the commercialization of its IoT platform; it sees applications in public utility services such as energy, water, and transportation. It also sees industrial applications where industry 4.0 makes IIoT applications to adapt to its IoT platform and utilizes it to provide smart industrial solutions. It also looks at creating a smart home that is connected to its IoT platform and gets various services such as healthcare, grocery management, transportation management, financial monitoring and management. The OceanConnect IoT platform talks about an Internet of Vehicles, which is a network of small vehicles with various IoT sensors in them that are capable of communicating with each other by sending data between them. Such a network of Internet of Vehicles will definitely help make traffic management easier and more real-time. However, in my opinion, the future will also lead to the creation of something known as the Internet of Devices, which will be the network of interconnected devices, not just limited to vehicles but to anything that uses IoT sensors, which are capable of getting onto any IoT platform with unique applications. It is this Internet of Devices, which is going to enable the smart life revolution period.

Another state-of-the-art implementation for IoT is the open connectivity foundation for all IoT connectivity. The problem that IoT is currently facing, which is going to become a major problem in the future, is that of common IoT standards in society. Without an IoT common standard, there will be various IoT devices and applications that will be isolated and not have the ability to talk to other IoT devices outside themselves. Nevertheless, there is a standard to do so now. There are three sets of people who are involved in the setting of standards for society. The first is the person developing the IoT application. The second is the set of consumers for whom the business is developing the IoT application. The third set is the developer or the creator of the IoT application who brings the IoT to life. This open connectivity foundation for IoT standards bridges the gap between all three. The key advantage is that there is a common definition of security interoperability and a common standard platform and an open source implementation and certification of IoT devices and applications. There are a lot of new models using this framework and they are getting released every month, such as smart pantry IoT devices, monitoring applications, and innovative applications such as smart doghouses, which the OCF currently supports. The complete ecosystem is being created by this Open IoT standard in order to make the IoT more secure and trustworthy.

Let's move on to the next section regarding a state-of-the-art implementation of machine learning and the IoT in the energy domain. We'll look at the key technological advancements that are enabling the energy sector to grow.

State-of-the-Art Implementation of Machine Learning and the IoT in the Energy Domain

The energy domain is going through a rapid technological wave right now all over the world. The heart of this change is the introduction of the IoT and machine learning for energy systems around the world. However, the adoption of the IoT and machine learning is not as straightforward as it is in the telecom sector. This is not just about creating new spectrum fields and using them. The energy sector has traditionally been a capital-intensive industry, be it oil, gas, hydro power, or even renewable energy sources such as solar or wind. This aspect of the energy domain makes it a slow adopter of technology such as the IoT and machine learning. The real reason is that most of the energy companies employ a lot of energy equipment and plants cost billions of dollars to commission and install.

The Current State of the IoT in the Energy Domain

The real challenge faced in this sector is about using the IoT and turning on old energy systems. It is easy to implement on energy equipment that has the capability to communicate using protocols such as Modbus interfaces. Very old equipment in the energy domain that does not have the capability to communicate using common industrial interfaces is definitely a big challenge for any energy company. It's not easy to replace some parts of a working mechanical plant in production because these energy systems can't have downtime. In such cases where the replacement of plant equipment cannot happen easily, the IoT and machine learning will have to use innovative ways in order to read data from such legacy equipment. After all, machine learning needs data in order to analyze and predict energy sector problems.

Solutions for Embracing the IoT in the Energy Domain

There are several companies around the world that are trying to develop solutions for this old equipment using new IoT sensor-based technology to capture data. The prominent one is Siemens (https://new.siemens.com/us/en/products/energy/featured-topics/redefine-performance.html). The key problems faced by the energy sector are the following:

1. Renewable energy plants are rising. This is happening in two ways. One is by old fossil fuel energy generation plants being replaced by solar or wind energy firms. The second is by the new energy plants being added to the grades which are renewable in nature.

    Decentralizing of energy operations is happening and this means that the power plant units are becoming more and more independent in order to generate energy. The old way of making energy, like oil and gas, required a centralized plant facility that would help create them. However, renewable energy generation systems like solar grids and wind grades allow the decentralization of energy generation. Decentralization allows the plans for energy generation to be operated in a small facility and also to make independent decisions

more conveniently for localized community
energy needs. Centralization gives the energy
companies the freedom to spread their operations
across geographies and use technology with the
IoT and machine learning in order to generate
energy. Decentralization does bring challenges
for making use of economies of scale; however,
it is compensated by the ability to make fast and
localized decisions on energy generation and the
flexibility that it brings along with it. The localized
small energy generation unit can make decisions
very quickly on when to cut the power generation or
up the power generation based on the consumption
pattern it observes using techniques like machine
learning or deep learning. The ability to adjust the
energy needs becomes the key differentiating factor
for establishing decentralized energy generation
units for the future.

2. We all know that the energy needs of humans,
   especially oil and gas, are increasing around
   the world and the power plants are being
   forced to perform outside of their original build
   specifications. This is a serious threat for the future
   as the capacity for the plans has already been
   breached and there is a stress on their operations
   for the future. There are various techniques that
   companies are adopting; some are using renewal
   energy plants beside the fossil fuel plants whereas
   others are increasing the capacity of existing fossil
   fuel plants.

3. A problem that most countries face is that of stabilization of the energy power grid and for that they need to have a good system that is able to predict energy needs well in advance for any part of the grade so that they can increase or decrease the energy generation before a spike or a bust happens. Blackouts can create problems for industries and citizens who are dependent on energy for their operations. So most countries are planning operations that will help them be more productive regarding the energy generation requirements of their communities. Having a stable grid for energy generation reduces the cost by providing a stable power economy versus if there is an unstable grade; the latter worsens the cost of energy generation and has a deeper economic impact on the community.

Siemens, a company that deals with the power plant and energy generation technologies, has conducted a survey of its clients on futureproofing the power plant with its energy domain. See https://assets.new.siemens.com/siemens/assets/public.1552332517.656a83b0-646a-4286-9546-dc54bd8a2e35.future-proofing-power-plant-ebook-2019.pdf. Customers in the survey talk about the future utility challenges that power plants anticipate facing in the next three years. The second problem mentioned by 30% of the respondents concerns adopting new AI technologies. This shows that the energy generation companies see challenges around adopting artificial intelligence machine learning and the IoT in their current environment. The outlook given in the survey was that of next three years, so it's a short-term outlook. In the question on success through planning and data analytics, a field that is closely associated with data science and machine learning, it asked its customers how successful plant modification and update projects

have been in the past three years; the responses from the companies that
are delivering plant operational objectives show that over 50% of them
met this target. However, the major point that comes out of the survey is
that the energy generators are investing heavily in planned upgrades and
modifications including new digital technologies, but the confidence levels
in deciding on modification and upgrades stands around 20 to 27% for
prioritizing it and 203% for deciding which partners should be involved
in such an upgrade project and between 18 to 33% for deciding when the
investment should be made or at what point of time that investment should
be made. The overall level of success reported by the upgrade projects in
the past three years is around 25%, which means that although upgrade
projects have been adopted by the energy generation companies, they have
met with very little success and there is a high failure rate.

So we have seen that the energy sector is in a state of flux, with many
companies being forced to upgrade their plans to more renewable energy
sources like solar and wind; however, the success rate of such project
upgrades is very low, leading to depletion in the return on investments.
The adoption of technologies like machine learning and artificial
intelligence is a challenge, as pointed out in the survey, and this means
that the sector is less prepared to adapt and use these technologies to
its advantage. But in any state-of-the-art implementation, a company
like Siemens provides solutions by using IoT sensors and devices and
connecting them to machine learning models in order to predict the
energy demands in any grid. This innovative solution can be applied
even to fossil fuel energy plants and to old data stores to analyze and
synthesize data to create prediction models. From predicting the periodic
maintenance cycles of plants and equipment to predicting energy spikes,
the machine learning models are being used by companies in order to
solve legacy problems (https://assets.new.siemens.com/siemens/
assets/public.1534921431.05f93d5a3c096441998512706a42840c51fd
3f68.2018-05-ew-article-iot-manfred-unterweger-en.pdf).

Another state-of-the-art implementation of energy is by a company named MindSphere, which specialises in digitization of the energy processes. It has evangelized the concept of using the Internet of Energy as an extension to the IoT by implementing Industry 4.0 where machines and processes are intelligently linked with each other so that they can work more efficiently and reliability through their entire service life. It has created the EN of the energy sector with a broad range of applications that are diverse in nature, from smart metering, digital services for protection, power except connectivity, power quality stability, critical power management systems, power plant condition monitoring systems, energy outage management systems, smart city platform systems, distribution of energy resource microgrid grades, and great prediction simulation models. These are some of the areas that are using smart IoT sensors to pick up a range of data, allowing business applications to evaluate the status of equipment and apply machine learning and data science on top for preventive maintenance and alert monitoring for circuit breakers, for example. MindSphere, a Siemens company, delivers the basis for successful digital transformation in the end and the realization of profits. It allows monitoring of field equipment and devices to make it possible to take appropriate actions as needed in an emergency situation in power plants. These kind of digital solutions are going to transform the implementation of the IoT and machine learning for the energy sector. The business activities of energy companies, which are very complex in nature and have criticality attached to them since they cater to many emergency services such as hospitals and airports, also make it very important for them to use artificial intelligence and machine learning in decision-making and planning assistance.

1. Complex operations distributed over the border
    involving exploration, generation, and distribution
    of energy requires the use of machine learning so
    that the data collected from all of these operations is
    analysed carefully by creating productive business
    models to aid the decision-making process of the
    energy sector management executives.

2. The high number and spread of customers pose
    challenges for energy companies in collecting
    data from the end user. The use of smart metering,
    which has the ability to communicate back data
    from individual customer devices such as smartgrid
    smartcloud storage applications and the use of
    machine learning to predict energy consumption
    based on this data can help the decision-making
    process in the energy sector.

3. Investments in infrastructures require major
    equipment and facilities and fleet upgrades or new
    plant commissioning and installation. The use of
    machine learning and AI for creating models before
    making such investments can greatly help in making
    the right decisions by providing data on what failed
    in the past and how to avoid the problem areas
    while implementing new upgrade or installation
    projects.

To improve operational performance, energy generation plants are
using the IoT for data by connecting their industrial assets to IoT sensor
devices and monitoring the performance of machinery at the plant level
and by measuring organizational KPI such as machinery productivity and
equipment efficiency. Predictive maintenance of old plants is definitely a

problem that IoT-based applications are solving for the energy sector. IoT data can be generated such that it makes them aware of the equipment performance payloads and locations of deteriorating energy assets. IoT-based pipeline energy meters have become very common; they use machine learning for measuring and predicting the flow of energy liquids like oil and gas. The quality of the flow can also be monitored and any deterioration can be predicted well in advance to avoid any huge losses during the production process. The energy sector is also using machine learning to preidct energy consumption in the fossil fuel plant so that it is in ready state to accommodate any spike in consumer energy requirement by having a standby renewable energy plant power up to take up the load. This balancing based on a predictive demand forecasting of energy requirements is an excellent use case for the energy sector.

You have looked at the state-of-the-art implementation of machine learning in the energy sector. The next section in this chapter covers the state-of-the-art implementation of machine learning and the IoT in the agriculture domain.

State-of-the-Art Implementation of Machine Learning and the IoT in the Agriculture Domain

The agriculture domain is seeing the emergence of an AI revolution right now. This technological wave is happening all over the world. The key transformation that is happening worldwide in farming is that major food conglomerates are trying to take over farming for their own consumption in producing products for the end consumer. The bigger food conglomerates want the farmers to produce specific patented varieties of crops in order to maintain good food quality and productivity. So in some countries they are tying up with farming communities and buying raw vegetables and beef from them. In other countries, they are hiring farmers on contract to produce specific varieties of food grains and other products such as potatoes, lettuce, carrots, dairy, and rapeseed, which are used as inputs for end consumer products such as burgers by McDonalds through the Flagship Farmers program (www.flagshipfarmers.com/en/about-the-program/).

The effort by these food giants is to create a bank of patented varieties of their core materials such as potatoes and lettuce. For example, McDonalds uses its famous patented Russet Burbank potato for its french fries (www.earthandtablelawreporter.com/2015/09/11/patenting-the-potato-not-all-taters-are-created-equal/). Another example is that of Monsanto, a food conglomerate that owns many patents for its vegetables and seeds, such as beans, broccoli, carrots, cucumber, and melons, among others. Many of these vegetables are commercially used by other companies. Monsanto's potatoes are used by PepsiCo for its flagship potato chip brand Lays (https://feast.media/food-brands-owned-by-monsanto). Monsanto holds 14 varieties of patents and the most popular is Roundup Ready® Corn 2 (www.monsantotechnology.com/content/genuity-traits-corn.aspx).

This is just an example of how food companies are buying a patent and utilizing it commercially. There is currently a controversy surrounding the use of GMO products as raw material for commercial foods. Genetically modified foods can cause various diseases, so they are not being used by major food companies; however, there is pressure on them to become commercially viable and to adapt to GMO-based practices to increase profits, which GMO food varieties offer.

The bold new alternative to genetically modified crops is computationally bred crop technologies combined with AI and machine learning to give excellent yields to farmers. The first one uses the IoT to gather data about the crop and its environment, such as the components of air using air filter sensors and oil sensors to gather data about the soil mix. The applications are already available through something known as statistical breeding patterns, which allow the farmer to get the best crop for the environment based on the atmosphere comprising of the air and soil mix. In this system, the farmer uses intelligent machine learning and IoT systems to select the best seed breed for the given air and soil mix based on past data in other regions of the world that have similar climates in order to produce maximum yields for their farms. This is being done by a company known as HFG; it is using farming data sets and predictive data science to build a predictive breeding platform for crops. (`www.zdnet.com/article/computational-breeding-can-ai-offer-an-alternative-to-genetically-modified-crops/`).

Another state-of-the-art implementation of AI in comes from technologically advanced nations like Japan for application in rice farming. People in rural areas are moving to cities to earn their livelihood. As a result of this mass migration of the population from rural areas to urban centers such as Tokyo and Kyoto, there are no people available to undertake farming in rural Japan. So Japan has an urgent need for farmers, especially for rice production, which is a staple. Automating and using machine learning and AI for farming in rural Japan can offer help. In this application, the farmers and developers in northeast Japan started using drones to supplement the workforce in the fields (see Figure 4-1).

Figure 4-1. *Drone robot for detecting disease onset in crops*

A drone named 91108 rich's was developed by a startup company. Its main function is to disperse pesticides and fertilizers on crops. It's able to do in 15 minutes work that would take a farmer 60 minutes. Control of such a drone through an iPad merged with a machine learning application to identify areas that require more fertilizer or pesticides can be developed on top of such an application. The diagnostic drones can detect onset of disease by continuous play Selma the crops and protecting any microorganisms near The Cross set on it pacific Best Buy all those areas are such an infestation occur. This is the where the use of IoT sensors on pest control along with machine learning applications running from nearby Raspberry Pi application station can help to automate the entire farming cycle (www.techrepublic.com/article/how-drones-are-changing-farming-in-rural-japan/?ftag=CMG-01-10aaa1b).

Summary

All of the applications you saw in this chapter are experimental in nature and have been newly introduced with the state-of-the-art implementation in the IoT of 5G by the Chinese company Huawei. You also saw how the challenges faced by energy sector are being resolved by the use of the IoT in areas of power generation both for fossil fuels and renewable energy sources. You then looked at the farming sector where the intelligent use of the IoT and machine learning promises to solve the human drain problem of farmers selling their land and migrating to the cities for a better living. In Chapter 5, you will prepare your setup for implementing case studies in these three domains.

CHAPTER 5

Preparing for the Case Studies Implementation

This chapter is all about preparing for the case studies implementation. You will first learn how to set up your Raspberry Pi 3 Model B+ with its hardware and software from scratch. Of course, you will be installing the popular Raspbian OS for this. After this, you will set up the Arduino Mega 2560 to your Raspberry Pi along with the IoT sensor modules to get data from. This will be followed by setting up the Python required for running a program on the Raspberry Pi. You will also be setting up the energy meter device with the Modbus enabled on it. After this, you will connect the Raspberry Pi with the Arduino and start communication between them. Lastly, you will test everything.

Setting Up Raspberry Pi 3 Model B+

Please refer back to Figure 3-1 in Chapter 3 for a schematic diagram on the components that you need to set up for the case studies. There are two software components, the Python IDE and the Arduino IDE, which you will set up once you have the hardware components fully set up. So let's get started with the hardware.

© Puneet Mathur 2020
P. Mathur, *IoT Machine Learning Applications in Telecom, Energy, and Agriculture*,
https://doi.org/10.1007/978-1-4842-5549-0_5

The first step to setting up your Raspberry Pi 3 Model B+ hardware is to decide on the operating system that you will use to run your applications. There are two popular operating systems available: Raspbian and Noobs. Noobs is used by people who are new to Raspberry Pi. You will be using the Linux version Raspbian, which is available from `www.raspberrypi.org/downloads/`. It can be seen in Figure 5-1, which shows the available operating system images for download from the official Raspberry Pi website.

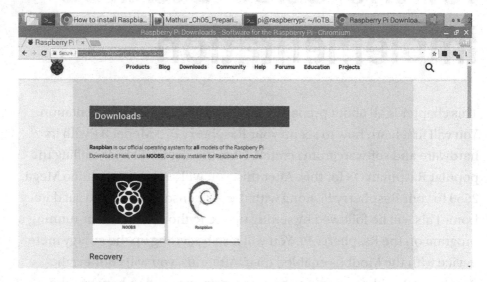

Figure 5-1. `www.raspberrypi.org` *download options*

Note that at this stage you have not booted up your Raspberry Pi 3 B+ as there is no operating system installed on its microSD card yet. A microSD card and microSD card reader or adapter come with the Raspberry Pi 3 B+ kit. If you do not have a microSD card, you can get one from the options on my website at `www.pmauthor.com/raspbian/`. For the installation of Raspbian, the base suggested microSD card size is 8GB. For Raspbian Lite image installations, Raspberrypi.org suggests at least 4GB. Do keep in mind that just the Raspberry Pi 3A+, 3B+, and Compute

Module 3+ can boot from an SD card bigger than 256GB. This is because there was a bug in the SoC utilized on past models of Pi. Also remember to note the microSD card class to which your microSD card belongs because that decides the supported writing speed for it; a class 4 card will most likely write at 4MB/s, while a class 10 ought to accomplish 10 MB/s. Do note this does not mean a class 10 card will beat a class 4 card for general read/write performance, in light of the fact that frequently this write speed is accomplished at the expense of read speed and a substantial increase seek times to the microSD card operations. The microSD card that comes with your Raspberry Pi 3 B+ kit should be latest and should most probably be class 10, as is common now; however, if you get a lower class like 4, you can order a faster card of class 10 at `www.pmauthor.com/raspbian/`.

Of course your laptop must have a slot for reading and writing to microSD cards either directly or through an adapter. This is shown in Figure 5-2. Also, you will need a computer or a laptop that is running either Windows or MacOS in order to download this image and then format the SD card and install this image on to it. I will show you how to do this on a Windows system in this chapter.

Figure 5-2. *The microSD card and microSD card adapter that comes with the official kit*

Do not insert the microSD card adapter into the Raspberry Pi 3 B+ board yet because it is empty and not formatted with any operating system image written on top of it yet. You need to first download the Raspbian image by clicking the download section of the website. Once you do so, you will get the option to download the latest image of Raspbian for your use. As of the writing of this book, the there are three options for download, as shown in Figure 5-3.

Figure 5-3. *Download options for Raspbian*

The first option is the Raspbian Buster, which is the name of the Raspbian flavor available for the desktop and recommended software. The second option is only the desktop and the third is the Lite version, which does not have all the features needed for you to do extensive programming for your IoT applications. So you will use the first option, which provides all the components of the software necessary to get going. You can click the Torrent or zip file depending on the software that you use to unzip it. I used the zip file version. The zip file for the Buster flavor of Raspbian is approximately 2.3GB so it takes time to download depending on your internet connection speed.

Once you have this downloaded, you need to insert your microSD card reader into the computer/laptop from which you are installing the operating system onto the SD card slot. Since you are using Windows, you will need a disk imaging utility. The free utility that I use is Win32 Disk Imager. If you do not have it, you can download it from this page: www.pmauthor.com/raspbian/. Once you have installed the Win32 Disk Imager utility after downloading it from the website, you can open it by double-clicking its icon. It should look like the one in Figure 5-4.

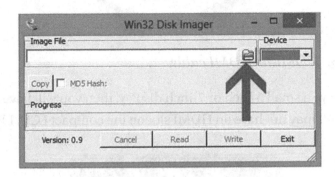

Figure 5-4. *Win32 Disk Imager*

Click the browse icon shown in the figure and locate your unzipped image file where you downloaded it. After selecting the location of the image file, click the Device drop-down box on the right-hand side and select your microSD Card drive letter from it. The Write button on the dialog box will get enabled, so click it to write the image. The progress bar will show you when it is done; this should not take more than a few minutes. Once it is done, click the Exit button to eject the microSD card. You now have an image of the latest Raspbian OS written on your microSD card and ready to boot. Before you can boot, however, you need to set up the SBC physical wires on the board and then attach the microSD card to the Raspberry Pi 3 B+ board.

The second step to setting up the Raspberry Pi 3 Model B+ is to get its board wired as shown in Chapter 2 with photos of the GPIO pins and the various components of the SBC. You need to plug in the bare minimum to get started. First, you need an LED/LCD TV or a display. There are many available, but the ones I prefer and use on my projects are listed at www.pmauthor.com/raspbian/. If you have a spare TV, you can use the HDMI-to-HDMI cable to connect it from your TV to the HDMI port on the SBC. See Figures 5-5 and 5-6.

Figure 5-5. *HDMI-to-HDMI cable*

If you have a small 5-inch or 7-inch display, then you will have a different cable and you may not have an HDMI slot on the compact LCD/LED panel.

Figure 5-6. *HDMI port on Raspberry Pi 3 B+*

The connected Raspberry Pi will look like the one in Figure 5-7. Make sure you get the connections right and do not to leave them loose in the port on both sides (the TV display and the Raspberry Pi 3 b+) or your display will not show up. This is a common beginner problem; after you power on your Raspberry Pi, you don't see the display coming up and you wonder what happened. It's an easy mistake to fix because the HDMI ports on the Raspberry Pi are quite large so you can see if the wire is secured properly on to the board. Also, make sure you do not power on the Raspberry Pi 3 B+ when you are fixing the HDMI cable onto it; this is important both as a safety precaution and as an operational requirement. Sometimes, when you have a powered-on Raspberry Pi and you fix the HDMI cable into its HDMI port, the display won't show up. In this case, restart the Raspberry Pi 3 B+ as there is no reboot button on this SBS.

There are two things I recommend doing before you finally boot up your Raspberry Pi. The first is to insert a USB mouse dongle into one of the four USB ports on the Raspberry Pi 3 B+ board. The second is to add another USB dongle for a wireless keyboard. These are essentials for you to control your SBC. You can see them connected to the SBC in Figure 5-7. Don't worry if you do not have a wireless version of a mouse or a keyboard; the wired versions of the USB mouse and keyboards work pretty well. It's just that the wired devices prove to be a bit too heavy for the SBC, which is very light and can move around if the wire of the mouse or keyboard shakes.

Figure 5-7. *USB dongles connected to the SBC*

Next, to get your Raspberry Pi to boot up, you need to plug it into a power source. The wonderful thing about this small SBC is that it uses microUSB as its power source. This means that you can power it via a power bank as well by attaching a microUSB power cable to it. Why do you need a power bank? It provides a power backup so that it does not shut down when the power goes off. This is important if you live in a region of the world where power cuts are frequent. If your region has stable power, then you do not need to connect to a power bank and you can connect it to the power source through a power adapter charger. You can see my connection to a power bank in Figure 5-8.

Figure 5-8. *Raspberry Pi 3 B+ power cord connected to a power bank*

What you see in the image is a Portronics 10000mAH power bank. It provides power for commercial applications for a few hours. In one of our factory energy audit applications, we had to create a power bank backup with one power bank connected to another one in serial, the first one feeding into the another one, so that even if the power went off for a few hours, the Raspberry Pi energy audit applications running on it would not be affected. Sometimes you need to think out of the box to build commercial grade applications that can take power cuts in its stride.

Now to bring your Raspberry Pi 3 B+ to life you just need to insert the microUSB side of the power cable into the power slot of the SBC, as shown in Figure 5-9. Please remember that if you have connected to a power source or a power bank and you are inserting the wire in the SBC, it will boot up in your hand. To prevent this, make sure the other end does not have power when you connect the microUSB to it.

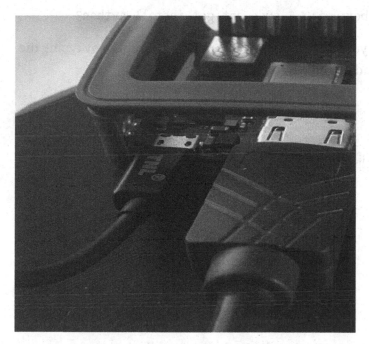

Figure 5-9. *microUSB and HDMI cables connected to Raspberry Pi 3 B+ SBC*

Now you are ready to boot up your device, so go ahead and switch on the power to your power adapter, which is connect to microUSB power cable. You will see a green LED blink near the HDMI port slot and the connected LED/LCD monitor should show the booting up image you saw in Chapter 3. Now you just have the basic Raspberry Pi with the latest version of Raspbian installed onto it. Next you need the software to be installed in order to work with the IoT sensors and devices.

If you are running Raspbian (not the Lite version), Python version 2.7 is preinstalled with the Raspbian distribution image. However, if you want to use Python 3, it needs to be installed through the following Raspbian command line prompt. Needless to say, you need an internet connection either through a connected Ethernet cable or a Wi-Fi connection to your Raspberry Pi.

```
pi@raspberrypi:~ $ sudo apt-get install python3
```

Test your Python 3 installation on the Raspbian by typing the command given in Figure 5-10.

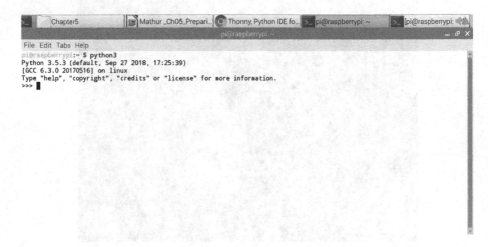

Figure 5-10. *Testing the Python 3 installation*

Now that you have the required Python version installed and tested, you need an integrated development environment (IDE) to program on Raspbian. I recommend you use the Thonny IDE because it lets you compile programs and run them in a single place inside an editor. I have discussed the need and features in Chapter 3 so let's get straight to the steps you need to install it on your Raspbian.

The Thonny IDE is free and comes preinstalled on most Raspbian OSes with Stretch program. If you do not find it in your installed programs, as you saw in Chapter 3, it can be downloaded from its website: `https://thonny.org/`. The good thing about this website is that it has a pictorial tutorial on its home page on how to use the IDE, which is pretty basic and simple. The installation can be done using the command in Listing 5-1.

Listing 5-1. Installing Thonny for Python 3

```
pi@raspberrypi:~ $ sudo apt-get install python3-thonny
Reading package lists... Done
Building dependency tree
Reading state information... Done
The following packages were automatically installed and are no
longer required:
  realpath vlc-plugin-notify vlc-plugin-samba vlc-plugin-video-
  splitter vlc-plugin-visualization
Use 'sudo apt autoremove' to remove them.
The following additional packages will be installed:
  python3-asttokens
Suggested packages:
  python3-distro
The following NEW packages will be installed:
  python3-asttokens
The following packages will be upgraded:
  python3-thonny
1 upgraded, 1 newly installed, 0 to remove and 205 not
upgraded.
Need to get 325 kB of archives.
After this operation, 146 kB of additional disk space will be used.
Do you want to continue? [Y/n]
y
0% [Connecting to archive.raspberrypi.org
(2a00:1098:0:80:1000:13:0:7)]
```

```
Get:1 http://archive.raspberrypi.org/debian stretch/ui armhf
python3-asttokens all 1.1.13-1+rpt1 [15.6 kB]
Get:2 http://archive.raspberrypi.org/debian stretch/ui armhf
python3-thonny all 3.1.0-1+rpt2 [309 kB]
Fetched 325 kB in 2min 2s (2,643 B/s)
Reading changelogs... Done
Selecting previously unselected package python3-asttokens.
(Reading database ... 149433 files and directories currently
installed.)
Preparing to unpack .../python3-asttokens_1.1.13-1+rpt1_all.deb
...
Unpacking python3-asttokens (1.1.13-1+rpt1) ...
Preparing to unpack .../python3-thonny_3.1.0-1+rpt2_all.deb ...
Unpacking python3-thonny (3.1.0-1+rpt2) over (3.0.5-1+rpt1) ...
Processing triggers for mime-support (3.60) ...
Processing triggers for desktop-file-utils (0.23-1) ...
Processing triggers for man-db (2.7.6.1-2) ...
Setting up python3-asttokens (1.1.13-1+rpt1) ...
Processing triggers for gnome-menus (3.13.3-9) ...
Processing triggers for hicolor-icon-theme (0.15-1) ...
Setting up python3-thonny (3.1.0-1+rpt2) ...
pi@raspberrypi:~ $
```

The Raspbian package for the Thonny IDE is downloaded from the
archives of the raspberry.org official website and then processed and
set up for use. If the installation is successful, you should see your IDE
installed under the Programming menu in the Start bar, as shown in
Figure 5-11.

Figure 5-11. *Verifying the Thonny Python IDE installation*

I discussed the verification part of the Thonny IDE in Chapter 3, so let's move to the next step of installing the required Python libraries for use in the case studies. The following is a list of Python libraries that you will need. Remember to change tkinter to Tkinter (yes, capital T), as discussed in Chapter 3.

- numpy is for operations on n-arrays and matrices in Python.

- pandas is used for data wrangling. It was designed for quick and easy data manipulation, aggregation, and visualization.

- Seaborn is for data visualization of statistical models through heatmaps and aggregrates.

- scikit-learn provides a consistent interface for machine learning algorithms used in your programs.

- Tkinter is for creating a graphical user interface for Python programs.

- scipy contains modules for linear algebra, optimization, integration, and statistics.

- matplotlib is for data visualization through various plots and graphs.

- NLTK is a natural language toolkit, and it used for common tasks of symbolic and statistical natural language processing.

- statsmodels is used to conduct data exploration through the use of various methods of estimation of statistical models and performing statistical assertions and analysis.

Now that you have seen the Python libraries that you need for the case studies, go ahead and install them. I created a script that does so on the Raspbian command line; see Figure 5-10. Please remember that the code is to install on Python 2.7; your need to change the script to add a 3 to pip and make it pip3 if you want to install all the libraries on Python 3.

Alert   If you use your own script for installing the Python libraries, make sure you use sudo before the pip install command, as shown in the script pipinstaller.run in Listing 5-2.

Listing 5-2. Raspbian Command Line Script to Install the Required Python Packages

```
pi@raspberrypi:~/$ cat pipinstaller.run
sudo pip install numpy | tee /home/pi/numpy.log

sudo pip install pandas | tee /home/pi/pandas.log

sudo pip install seaborn | tee /home/pi/seaborn.log

sudo pip install sklearn | tee /home/pi/sklearn.log
```

```
sudo pip install tkinter | tee /home/pi/tkinter.log

sudo pip install scipy | tee /home/pi/scipy.log

sudo pip install nltk | tee /home/pi/nltk.log

sudo pip install statsmodels | tee /home/pi/statsmodels.log
pi@raspberrypi:~/
```

You will notice that I appended sudo before every command and then piped the output to the command to log the output into a file for each of the Python libraries. So for sudo pip install scipy | tee /home/pi/ scipy.log, the output is stored in a separate log file named scipy.log. It's easier to be able to look at each of the log files separately and to debug and see if there were any errors during installation. You may ask why am I storing all these outputs in a log file and the simple answer is because on Raspbian 3 Model b+ it can take several hours for a package to install depending upon its size and the download speed of the Internet in your area. So keep in mind this important point before running the script and make sure you have power backup running for your Raspberry Pi system and display. Everything should have backup power before you attempt to run this script. The backup power for your monitor is equally important because the Raspberry Pi Model 3 B+ fails to show the screen sometimes if your monitor is switched off for some reason, like a power cut. The output of the script may look like Listing 5-3.

Listing 5-3. Output of the Pip Python Package Installer Script

```
Requirement already satisfied: numpy in /usr/lib/python2.7/
dist-packages
Requirement already satisfied: pandas in /usr/lib/python2.7/
dist-packages
Requirement already satisfied: numpy>=1.7.0 in /usr/lib/
python2.7/dist-packages (from pandas)
```

Requirement already satisfied: python-dateutil in /usr/lib/
python2.7/dist-packages (from pandas)
Requirement already satisfied: pytz>=2011k in /usr/lib/
python2.7/dist-packages (from pandas)
Requirement already satisfied: seaborn in /usr/lib/python2.7/
dist-packages
Requirement already satisfied: sklearn in /usr/local/lib/
python2.7/dist-packages
Requirement already satisfied: scikit-learn in /usr/lib/
python2.7/dist-packages (from sklearn)
Collecting tkinter
  Could not find a version that satisfies the requirement
tkinter (from versions:)
No matching distribution found for tkinter
Requirement already satisfied: scipy in /usr/lib/python2.7/
dist-packages
Collecting nltk
  Downloading https://files.pythonhosted.org/packages/87/16/4d2
47e27c55a7b6412e7c4c86f2500ae61afcbf5932b9e3491f8462f8d9e/nltk--
3.4.4.zip (1.5MB)
Collecting singledispatch; python_version < "3.4" (from nltk)
  Downloading https://files.pythonhosted.org/packages/c5/10/3
69f50bcd4621b263927b0a1519987a04383d4a98fb10438042ad410cf88/
singledispatch-3.4.0.3-py2.py3-none-any.whl
Requirement already satisfied: six in /usr/lib/python2.7/dist-
packages (from nltk)
Building wheels for collected packages: nltk
  Running setup.py bdist_wheel for nltk: started
  Running setup.py bdist_wheel for nltk: finished with status
'done'
  Stored in directory: /root/.cache/pip/wheels/41/c8/31/48ace44
68e236e0e8435f30d33e43df48594e4d53e367cf061

Successfully built nltk
Installing collected packages: singledispatch, nltk
Successfully installed nltk-3.4.4 singledispatch-3.4.0.3
Requirement already satisfied: statsmodels in /usr/lib/
python2.7/dist-packages

In my case, as you can see from the installer log output, I had all the necessary libraries except for nltk, in which case it was downloaded and installed. To verify if there were any errors during setup of nltk library, the output of nltk.log file is shown in Listing 5-4.

Listing 5-4. Output of the nltk.log File

Collecting nltk
  Downloading https://files.pythonhosted.org/packages/87/16/4d2
47e27c55a7b6412e7c4c86f2500ae61afcbf5932b9e3491f8462f8d9e/nltk--
3.4.4.zip (1.5MB)
Collecting singledispatch; python_version < "3.4" (from nltk)
  Downloading https://files.pythonhosted.org/packages/c5/10/3
69f50bcd4621b263927b0a1519987a04383d4a98fb10438042ad410cf88/
singledispatch-3.4.0.3-py2.py3-none-any.whl
Requirement already satisfied: six in /usr/lib/python2.7/dist-
packages (from nltk)
Building wheels for collected packages: nltk
  Running setup.py bdist_wheel for nltk: started
  Running setup.py bdist_wheel for nltk: finished with status
'done'
  Stored in directory: /root/.cache/pip/wheels/41/c8/31/48ace44
68e236e0e8435f30d33e43df48594e4d53e367cf061
Successfully built nltk
Installing collected packages: singledispatch, nltk
Successfully installed nltk-3.4.4 singledispatch-3.4.0.3

So no errors were found in this file and the python nltk library was installed successfully. You may want to check all the other log files that were installed for you one by one, like `scipy.log`, `statsmodels.log`, etc. Now you have successfully installed all the required Python libraries.

Time Out for Testing

I asked you at the beginning of Chapter 3 to not execute any code and enjoy the execution of the IoT with machine learning along with Arduino. But here you are setting up all the infrastructure software and hardware that you need to get up and running for the case studies ahead.

However, before proceeding to the execution in the next chapter, I urge you to go back to Chapter 3 and execute all the code for testing your Python libraries given in Figure 3-3, the Hello world Python code that tests the installed libraries on your Raspbian. Figure 3-4 shows the output of the code on the Thonny IDE and you should get a similar output too. If not, you may have not installed your Python libraries correctly; you may need to go back and check your installation logs from Figure 5-11 to check where the error occurred. Then you should go ahead and execute the code for testing the Pandas dataframe in Figure 3-5, whose output should be similar to that shown in Figure 3-6 on the Thonny IDE. Further, you should test the small code in Figure 3-7 for the sklearn machine learning integrated library, the output of which appears in Figure 3-8 of Chapter 3. These are tests for basic libraries; however you could check the advanced libraries like tkinter, gpiozero, smtplib, pygame, psutil, platform, and time, which were used in the Python code of Chapter 3 via the Raspbian command line script given in Listing 5-5, the `iotpypkgtest.run` program. I showed a permission denied error that you may get on the command line the first time you execute it; it can be solved by changing the permissions using the `chmod 700` command for the script.

Listing 5-5. Script for Testing the Advanced Packages Installation

```
pi@raspberrypi:~/$ ./iotpypkgtest.run
bash: ./iotpypkgtest.run: Permission denied
pi@raspberrypi:~/$ chmod 700 iotpypkgtest.run
pi@raspberrypi:~/$./iotpypkgtest.run
pi@raspberrypi:~/$
pi@raspberrypi:~/$ ls -l *.out
-rw-r--r-- 1 pi pi   0 Aug 17 23:19 gpiozero.out
-rw-r--r-- 1 pi pi   0 Aug 17 23:19 platform.out
-rw-r--r-- 1 pi pi   0 Aug 17 23:19 psutil.out
-rw-r--r-- 1 pi pi   0 Aug 17 23:19 pygame.out
-rw-r--r-- 1 pi pi   0 Aug 17 23:19 smtplib.out
-rw-r--r-- 1 pi pi   0 Aug 17 23:19 time.out
-rw-r--r-- 1 pi pi 111 Aug 17 23:19 tkinter.out
pi@raspberrypi:~/$
```

I deliberately used the command python -c 'import tkinter' in the script to create an error so that you can see in Listing 5-5 the rest of the files are of zero bytes; however, the tkinter.out file has 111 bytes written to it, signaling that there is some error there. You saw the reason for the error in Chapter 3; however, to reiterate, the tkinter library is used for GUI creation purposes such as window screens and buttons to manipulate the IoT devices and works with Python 3.x instead of Python 2.7. On your Raspbian command line, when you type just python, you are invoking the Python 2.7 interpreter; if you want to invoke the Python version 3 interpreter, you need to type python3 in the command line, like I have done in Listing 5-6.

Listing 5-6. Successful Execution of tkinter on Python 3.x Intepretor

```
pi@raspberrypi:~/$ python3 -c "import tkinter"
pi@raspberrypi:~/$
```

The error from the Python 2.7 command line is shown in Listing 5-7.

Listing 5-7. Error from Python 2.7 Command Line

```
pi@raspberrypi:~/$ cat tkinter.out
Traceback (most recent call last):
  File "<string>", line 1, in <module>
ImportError: No module named tkinter
pi@raspberrypi:~/$
```

The error is not very intuitive but it does say there is no module named tkinter that this version of the interpreter recognizes. You may get a different error if there is a problem with any of the other Python library installations and that will need to be debugged by you separately. So far you have installed the basic libraries necessary for Python programs to run including sklearn for machine learning algorithms. However, for IoT to work through Raspberry Pi 3 B+ and Arduino Mega 4560, you need the advanced libraries from the following list:

- gpiozero: Main library that enables communication between Python and the GPIO ports on Raspberry Pi 3 B+.

- smtplib: Used to send email alerts from the Raspbian Linux-based OS.

- pygame: Used to generate sounds in your Python programs.

- psutil (process and system utilities): A cross-platform library for retrieving information on **running processes** and **system utilization** (CPU, memory, disks, network, sensors) in Python.

- platform: Used to access the underlying platform's data, such as hardware, OS, and interpreter version information.

- time: Python has defined a package named time which allows you to handle various operations regarding time and its conversions and representations, which find use in your applications for storing the timestamp of an event through your IoT devices.

The gpiozero is the main python package I use on Raspberry Pi 3 B+, which has been tested and run successfully on commercial applications for my clients, so I recommend using this package for communicating with IoT devices. What is the use of having an application that does monitoring and alerting but does not communicate in real time to the required people? For this you need the smtplib python library. Configuring a SMTP server on your Raspberry Pi 3 B+ is beyond the scope of this book so I will not cover this part. However, once you have it configured, it is easy to send an email using the machinemon applications given in Chapter 3. The psutil and platform Python packages are used to get data from the underlying hardware of Raspberry Pi 3 B+ such as CPU percentage, temperature, memory, disk space, etc. To create a successful commercial application, you need to store a timestamp of every event that occurs during monitoring through IoT devices and for this the time Python package comes in handy.

You now know about the advanced packages that are to be used for IoT, so go ahead and install them using a script named iotpypackagesinstall.run and a test named iotpypackages.run, shown in Listings 5-8 and 5-9, respectively.

Listing 5-8. Code to Run iotpypackages.run Raspbian Script

```
sudo pip install gpiozero | tee gpiozero.log
sudo pip install pygame | tee  pygame.log
sudo pip install psutil | tee  psutil.log
pi@raspberrypi:~$ ./iotpypackagesinstall.run
```

```
Requirement already satisfied: gpiozero in /usr/lib/python2.7/
dist-packages
Requirement already satisfied: pygame in /usr/lib/python2.7/
dist-packages
Requirement already satisfied: psutil in /usr/lib/python2.7/
dist-packages
#Testing installed packages using script
pi@raspberrypi:~/IoTBook/Chapter5 $ ./iotpypackages.run
```

Listing 5-9. Output of Script iotpypackages.run

```
pi@raspberrypi:~/IoTBook/Chapter5 $ cat iotpypackages.run
python -c "import tkinter" &> tkinter.out
python -c "import gpiozero" &> gpiozero.out
python -c "import smtplib" &> smtplib.out
python -c "import pygame" &> pygame.out
python -c "import psutil" &> psutil.out
python -c "import platform" &> platform.out
python -c "import time" &> time.out
pi@raspberrypi:~/IoTBook/Chapter5 $ ls -l *.out
-rw-r--r-- 1 pi pi   0 Aug 18 00:51 gpiozero.out
-rw-r--r-- 1 pi pi   0 Aug 18 00:51 platform.out
-rw-r--r-- 1 pi pi   0 Aug 18 00:51 psutil.out
-rw-r--r-- 1 pi pi   0 Aug 18 00:51 pygame.out
-rw-r--r-- 1 pi pi   0 Aug 18 00:51 smtplib.out
-rw-r--r-- 1 pi pi   0 Aug 18 00:51 time.out
-rw-r--r-- 1 pi pi 111 Aug 18 00:51 tkinter.out
pi@raspberrypi:~/IoTBook/Chapter5 $
```

As you can see from the test script output in Listing 5-9, all the files are 0 bytes except for tkinter.out. So the advanced packages have been installed successfully. If you find any issues during these Python library

installations, you can write to me through the technical forums at
www.pmauthor.com/raspbian/ and I will try to help resolve your queries.

So far you have successfully set up your Raspberry Pi 3 B+ and also
Python with its required packages and tested them through scripts in
this chapter and Chapter 3. Now you must install the Arduino IDE to
communicate with the Arduino Mega 4560 and create a SQLite3 database
where your IoT applications can to store data. After installing the Arduino
IDE, you will need to test communication between Raspberry Pi and
Arduino using a serial cable. You will look at the detailed steps of setting
up the hardware of the Arduino Mega 4560 to Raspberry Pi 3 B+ in the next
steps.

Determining Bit Size

Before you proceed to installing the Arduino IDE, you need to know
whether your Raspberry Pi is 32-bit or 64-bit. For this, there are two useful
commands that are given in Listing 5-10, which provide the bash bit
version.

Listing 5-10. Finding Out the Bit Version of Your Raspberry Pi

```
> which bash
/bin/bash
> file /bin/bash
/bin/bash: ELF 32-bit LSB executable, ARM, version 1 (SYSV) ...
```

My Raspberry Pi 3 B+ has a 32 bit bash executable. This is not a very
definitive way of determining your bit version, but it is most unlikely that
a 32-bit bash has been compiled on a 64-bit processor by the hardware
manufacturer. Further, to confirm, you can use the uname command
shown in Listing 5-11, which shows that I am running a Raspberry Pi Linux
variant on an ARMv7 processor.

Listing 5-11. Command for Confirming the Processor Information

```
pi@raspberrypi:~/$ uname -a
Linux raspberrypi 4.14.98-v7+ #1200 SMP Tue Feb 12 20:27:48 GMT
2019 armv7l GNU/Linux
```

Even this does not clearly say the width bit of the processor. It is important to know this information because you don't want to install the wrong Arduino bit IDE and mess up your Raspbian installation. So you can use a program known as lshw, which you can install using the command shown in Listing 5-12.

Listing 5-12. Installing the lshw Utility to Find Out the Processor Bit

```
pi@raspberrypi:~/$ sudo apt-get install lshw
Reading package lists... Done
Building dependency tree
Reading state information... Done
The following packages were automatically installed and are no
longer required:
  realpath vlc-plugin-notify vlc-plugin-samba vlc-plugin-video-
  splitter
  vlc-plugin-visualization
Use 'sudo apt autoremove' to remove them.
The following additional packages will be installed:
  libpci3 pciutils
The following NEW packages will be installed:
  libpci3 lshw pciutils
0 upgraded, 3 newly installed, 0 to remove and 205 not upgraded.
Need to get 525 kB of archives.
After this operation, 1,879 kB of additional disk space will be
used.
```

```
Do you want to continue? [Y/n] y
Get:1 http://raspbian.mirror.net.in/raspbian/raspbian stretch/
main armhf libpci3 armhf 1:3.5.2-1 [50.9 kB]
Get:2 http://raspbian.mirror.net.in/raspbian/raspbian stretch/
main armhf pciutils armhf 1:3.5.2-1 [271 kB]
Get:3 http://raspbian.mirror.net.in/raspbian/raspbian stretch/
main armhf lshw armhf 02.18-0.1 [203 kB]
Fetched 525 kB in 2s (188 kB/s)
Selecting previously unselected package libpci3:armhf.
(Reading database ... 149467 files and directories currently
installed.)
Preparing to unpack .../libpci3_1%3a3.5.2-1_armhf.deb ...
Unpacking libpci3:armhf (1:3.5.2-1) ...
Selecting previously unselected package pciutils.
Preparing to unpack .../pciutils_1%3a3.5.2-1_armhf.deb ...
Unpacking pciutils (1:3.5.2-1) ...
Selecting previously unselected package lshw.
Preparing to unpack .../lshw_02.18-0.1_armhf.deb ...
Unpacking lshw (02.18-0.1) ...
Setting up lshw (02.18-0.1) ...
Processing triggers for libc-bin (2.24-11+deb9u3) ...
Processing triggers for man-db (2.7.6.1-2) ...
Setting up libpci3:armhf (1:3.5.2-1) ...
Setting up pciutils (1:3.5.2-1) ...
Processing triggers for libc-bin (2.24-11+deb9u3) ...
```

Now that you have this utility installed, you can type the command
lshw. The output in Listing 5-13 clearly gives shows the width bit as 32-bit
for my processor.

Listing 5-13. Output of the lshw Command

```
pi@raspberrypi:~/$ sudo lshw
USB
raspberrypi
    description: ARMv7 Processor rev 4 (v7l)
    product: Raspberry Pi 3 Model B Plus Rev 1.3
    serial: 0000000042f31c73
    width: 32 bits
    capabilities: smp
  *-core
      description: Motherboard
      physical id: 0
    *-cpu:0
        description: CPU
        product: cpu
        physical id: 0
        bus info: cpu@0
        size: 1400MHz
        capacity: 1400MHz
        capabilities: half thumb fastmult vfp edsp neon vfpv3
        tls vfpv4 idiva idivt vfpd32 lpae evtstrm crc32
        cpufreq
    *-cpu:1
        description: CPU
        product: cpu
        physical id: 1
        bus info: cpu@1
        size: 1400MHz
        capacity: 1400MHz
        capabilities: half thumb fastmult vfp edsp neon vfpv3
        tls vfpv4 idiva idivt vfpd32 lpae evtstrm crc32 cpufreq
```

```
*-cpu:2
     description: CPU
     product: cpu
     physical id: 2
     bus info: cpu@2
     size: 1400MHz
     capacity: 1400MHz
     capabilities: half thumb fastmult vfp edsp neon vfpv3
     tls vfpv4 idiva idivt vfpd32 lpae evtstrm crc32 cpufreq
*-cpu:3
     description: CPU
     product: cpu
     physical id: 3
     bus info: cpu@3
     size: 1400MHz
     capacity: 1400MHz
     capabilities: half thumb fastmult vfp edsp neon vfpv3
     tls vfpv4 idiva idivt vfpd32 lpae evtstrm crc32 cpufreq
*-memory
     description: System memory
     physical id: 4
     size: 927MiB
*-usbhost
    product: DWC OTG Controller
    vendor: Linux 4.14.98-v7+ dwc_otg_hcd
    physical id: 1
    bus info: usb@1
    logical name: usb1
    version: 4.14
    capabilities: usb-2.00
    configuration: driver=hub slots=1 speed=480Mbit/s
```

```
*-usb
      description: USB hub
      product: USB 2.0 Hub
      vendor: Standard Microsystems Corp.
      physical id: 1
      bus info: usb@1:1
      version: b.b3
      capabilities: usb-2.00
      configuration: driver=hub maxpower=2mA slots=4
      speed=480Mbit/s
    *-usb:0
          description: USB hub
          product: USB 2.0 Hub
          vendor: Standard Microsystems Corp.
          physical id: 1
          bus info: usb@1:1.1
          version: b.b3
          capabilities: usb-2.00
          configuration: driver=hub maxpower=2mA slots=3
          speed=480Mbit/s
        *-usb:0
              description: Generic USB device
              vendor: Standard Microsystems Corp.
              physical id: 1
              bus info: usb@1:1.1.1
              version: 3.00
              capabilities: usb-2.10
              configuration: driver=lan78xx maxpower=2mA
              speed=480Mbit/s
        *-usb:1
              description: Mouse
              product: 2.4G Mouse
```

```
         vendor: Telink
         physical id: 2
         bus info: usb@1:1.1.2
         version: 1.00
         capabilities: usb-1.10
         configuration: driver=usbhid maxpower=50mA
         speed=12Mbit/s
    *-usb:2
         description: Keyboard
         product: USB Receiver
         vendor: Logitech
         physical id: 3
         bus info: usb@1:1.1.3
         version: 24.07
         capabilities: usb-2.00
         configuration: driver=usbhid maxpower=98mA
         speed=12Mbit/s
   *-usb:1
         description: Communication device
         product: Mega 2560 R3 (CDC ACM)
         vendor: Arduino (www.arduino.cc)
         physical id: 2
         bus info: usb@1:1.2
         version: 0.01
         serial: 55739323737351F052A1
         capabilities: usb-1.10
         configuration: driver=cdc_acm maxpower=100mA
         speed=12Mbit/s
*-network:0
      description: Wireless interface
      physical id: 2
      logical name: wlan0
```

```
        serial: b8:27:eb:a6:49:26
        capabilities: ethernet physical wireless
        configuration: broadcast=yes driver=brcmfmac
        driverversion=7.45.154 firmware=01-4fbe0b04
        ip=172.20.10.11 multicast=yes wireless=IEEE 802.11
   *-network:1
        description: Ethernet interface
        physical id: 3
        logical name: eth0
        serial: b8:27:eb:f3:1c:73
        size: 10Mbit/s
        capacity: 1Gbit/s
        capabilities: ethernet physical tp mii 10bt 10bt-fd
        100bt 100bt-fd 1000bt-fd autonegotiation
        configuration: autonegotiation=on broadcast=yes
        driver=lan78xx driverversion=1.0.6 duplex=half link=no
        multicast=yes port=MII speed=10Mbit/s
pi@raspberrypi:~/$
```

While this command gives an awful lot of information about the hardware, the information we are interested in is given in the sixth line of the command output: **width: 32 bits**. This confirms in a definitive manner that my Raspberry Pi 3 B+ is running on a 32-bit processor.

Installing the Arduino IDE

Since you know your processor bit width, you can move to the actual installation of the Arduino IDE, which can be done in two ways. One way is to go to www.arduino.cc and click Software ➤ Downloads, as shown in Figure 5-12.

Figure 5-12. *Downloading the Arduino IDE from* www.arduino.cc

Now scroll down to the Arduino IDE download section shown in Figure 5-13 and choose the operating system with the corresponding bit (32 or 64) for download.

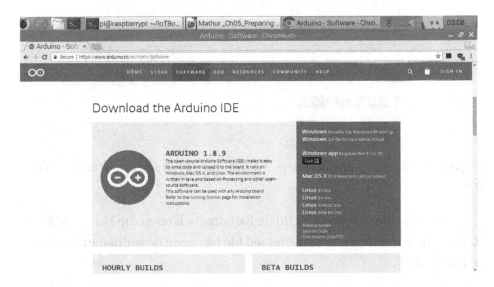

Figure 5-13. *Download the Arduino IDE*

149

As you can see, there are various options in the blue bar on the right-hand side of the page. Linux has four options: Linux 32 bits, Linux 64 bit, Linux ARM 32 bits, and Linux ARM 64 bits. The uname and lshw commands showed that I am running Linux on an ARM v7 processor with a width bit of 32 bits, so I select Linux ARM 32 bits for download in my case. If you have a 64-bit processor, you need to download the Linux ARM 64-bit IDE. Click the Just Download button on the next page if you do not want to contribute a small donation towards Arduino development.

After the download ends, follow the steps given in Figure 5-14 by clicking Show in Folder option in your browser.

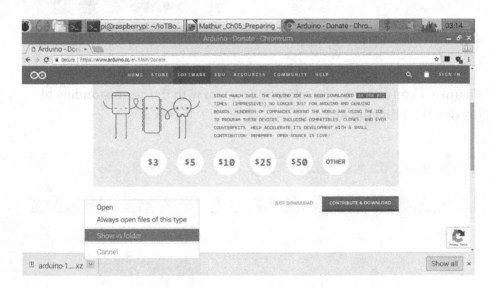

Figure 5-14. *Click the Show in Folder option to see the downloaded file*

After this, the Raspbian GUI File Explorer will open the Downloads folder where the Arduino compressed file has been downloaded, as shown in Figure 5-15.

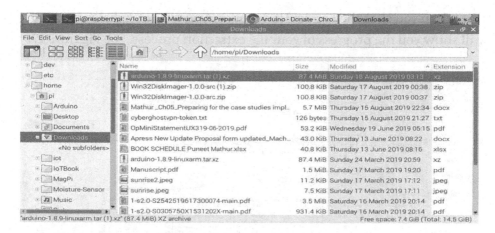

Figure 5-15. *Downloads folder with the *.xz compressed file*

You need to extract the tar file from inside the ∗.xz extension file, as shown in Figure 5-16. Right-click the file to select and left-click the Extract Here menu option.

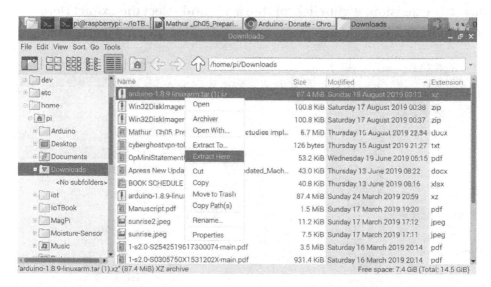

Figure 5-16. *The Extract Here option for the *.xz file*

151

When you click the Extract Here option, you will see a dialog box that will show you the progress of your file extraction process. See Figure 5-17.

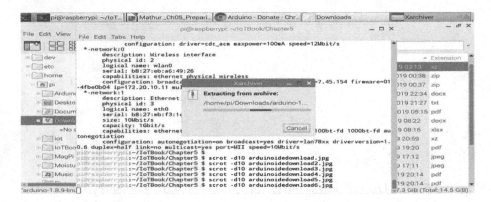

Figure 5-17. *Extraction in progress for the *.xz file*

Once you have this extracted, you will be able to see a new file created in the Downloads folder with the same name as the *.xz file but now with a *.tar extension. Right-click this file again and select the Extract Here option from the dropdown menu once again, as shown in Figure 5-18.

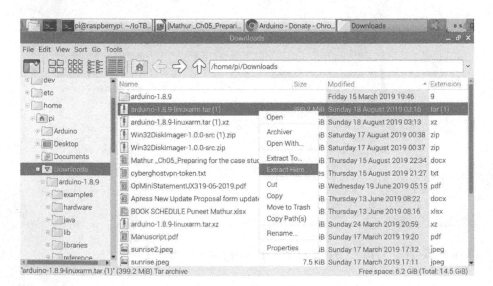

Figure 5-18. *Extraction of the tar file*

After the extraction process is complete, you will now see a new folder created, as shown in Figure 5-19.

Figure 5-19. *New Arduino folder after extraction*

You can go into the folder by double-clicking it, as shown in Figure 5-20. You will see a file named `install.sh`. Double-click this file and the Arduino IDE will be installed on your system.

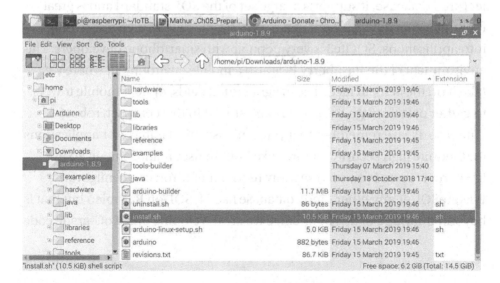

Figure 5-20. *Running the install.sh file to finish the installation*

The second, easier method to install the Arduino IDE just involves executing three commands:

```
sudo apt-get update
sudo apt-get upgrade
sudo apt-get install arduino
```

Although the command line method of installing the Arduino IDE is much simpler and has lot less steps, some people like the GUI more than the command line, so I have shown both the methods and leave it up to you on how you want to install it. Now you are at the data storage step in your setup, that of installing a SQLite3 database on Raspbian.

Installing SQLite3 Database

SQLite3 is a free database that you are going to use easily to create and use a database in your applications for storage. Although SQLite3 is not a full-featured database, it supports a large set of the SQL standard and is great for application development that need a simple database engine to plug into applications. SQLite3 is very popular with smartphone developers. The environment of the Raspberry Pi is also a mobile environment where we take it on the field, in factories, in agricultural fields, or near mobile towers to gather data through IoT sensors and store it in this compact, robust database. Please remember that you can't use this database for applications that require high security features like built-in user management. So use it wherever there is no such security requirement. You can configure PostgreSQL or any other Linux database like MySQL on Raspbian but that is beyond the scope of this book since we are developing only PoC-grade code.

Use the following script to install Sqlite3 onto the Raspbian OS:

```
sudo apt-get update
sudo apt-get upgrade
sudo apt-get install sqlite3
sqlite3 <firstdb.db>
SQLite version 3.8.7.1 2014-10-29 13:59:56

Enter ".help" for usage hints.
sqlite>
```

You can also install SQLite3 in the manner I showed you for the GUI installation for the Arduino IDE from the official website of the SQLite organization at www.sqlite.org/download.html.

Now you've come to the last step of installing the Modbus interface device, the single phase meter, onto the Arduino.

Disclaimer   Before proceeding further, I would like to warn you of the risk associated with using this Modbus device. The energy meter is an electrical appliance and carries the risk of short-circuiting if not connected properly. It can also be dangerous to human life if not used as per its instruction manual. It can also burn and damage your Raspberry Pi and Arduino boards if wrong connections are made. So if you are not comfortable with electric connections, I strongly advise you to not use this device or to get expert help from a local electrical technician to make the proper connections. Neither I nor the publisher can be held responsible in any way whatsoever for any kind of damages, either material or to life. User discretion is highly advised.

The additional devices that you will be using to set up the energy meter are shown in Figures 5-21 through 5-23:

1. Schneider Electric Conzerv EM6400 Series Power Meter

Figure 5-21. *Conzerv EM6400 Energy Meter*

2. 5V RS485 to TTL Signal Mutual Conversion Module for Arduino Overvoltage Protection with Signal Indicator

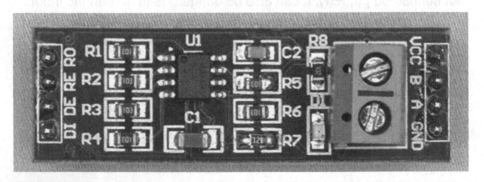

Figure 5-22. *5V RS485 to TTL Signal Mutual Conversion Module for Arduino*

Figure 5-23. *Connected 5V Modbus Conversion Module*

Let's now connect the wires between the Arduino Mega 256 and the 5V RS485 Modbus Conversion Module. I have shown the wires attached to the Modbus Conversion Module in the Arduino Mega 2560.

In order to connect the Modbus Conversion Module to Arduino, you need to first understand the layout of the Arduino Mega 2560 PCB, which is shown in Figure 5-24. Notice in the diagram that the upper pins pertain to communication and the GPIO pins. The right side of the board has the data pins and the bottom part of the board has the power and analog pins. This board is very different from the Arduino Uno board because it is used for controlling multiple slaves and is used in industrial applications. This is why there are so many data pins in the PCB layout.

Now connect the Modbus Conversion Module to the Arduino Mega board. You have to connect three pins to the power area of the board, which is near the power controller on the bottom left-hand side of the layout diagram in Figure 5-24. The left-most dark blue pin gets connected to the pin that has 5V written under it. The next green wire is connected to the immediate next pin marked as GND. Ignore the yellow wire on the left of Figure 5-24; you don't need it for your setup. You can see this in Figure 5-24—the three pins connected to the power section of the Arduino Mega board. Please remember these are male jumper wires that you need because the Arduino Mega board only has holes and not pins like the Raspberry Pi GPIO setup.

157

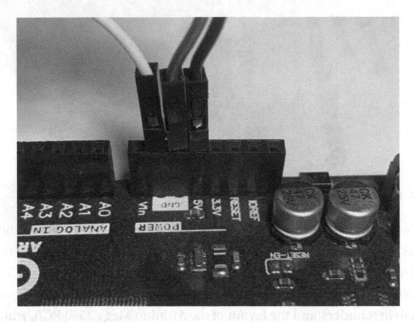

Figure 5-24. *Modbus Conversion Module wire connections to Arduino Mega board*

The wires on the other end have to be female jumper wires because the Modbus Conversion Module has pins to put the wires into. So you should have two male-to-female jumper wires for this connection. Connect the other end of these two wires to the Modbus Conversion Module pins marked as VCC and GND (VCC is the 5V power wire and GND is the grounding of the circuit). You can refer to Figure 5-25.

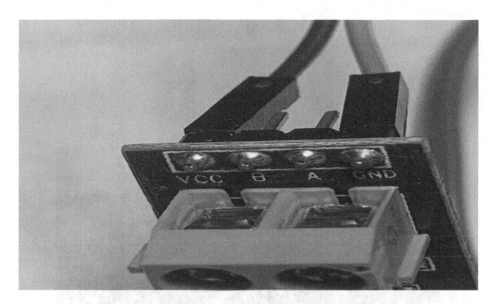

Figure 5-25. *Power and ground connections to the Modbus Conversion Module*

The 5V dark blue wire connected on Arduino connects to the VCC pin marked on the Modbus Conversion Module. The GND green wire pin on the Arduino board connects to GND on the Modbus Conversion Module PCB.

Now one part of your Modbus Conversion Module is connected to the Arduino board. You need to connect the other end to the GPIO pins section of the Arduino board or the communication side for a total of four pins. See Figure 5-26. On the Arduino board

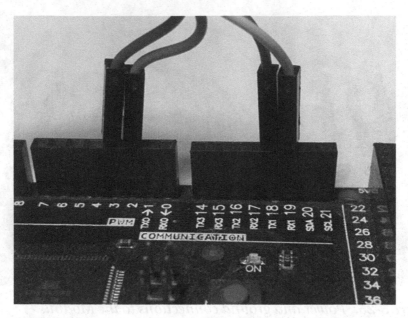

Figure 5-26. *Connections to the PWM and communications side of Arduino Mega board*

As you can see, there are two pins on the left-hand side of the board named PWM connected to pins numbered 3 and 2, which are adjacent to each other. The next two pins are connected to the communications pins 18 and 19, which are marked as TX1 and RX1, respectively, on the Arduino Mega board. Once you have this in place, you need to connect the other end of the four pins in the following manner. Pin 1 connected on the Arduino board to pin number 3 of PWM needs to connect to the pin marked DE on the Modbus Conversion Module pin. Pin 2 connected on the Arduino board to pin number 2 of PWM needs to connect to the pin marked RE in the middle of the Modbus Conversion Module. Similarly, pin 3 connected on the Arduino board to pin number 18 in Figure 5-27 needs to connect to the pin marked DI on the Modbus Conversion Module. The fourth pin connected to pin number 19 on the Arduino board should be connected to the pin marked RO on the Modbus Conversion Module. This completes the wiring process for the Modbus Conversion Module to the Arduino Mega board. You can check your connections against Figure 5-27.

Figure 5-27. *Wiring to the Modbus Conversion Module communication and PWM pins*

Now you are ready for the next step: to connect your Arduino module to the Raspberry Pi board. For this, you need to connect a serial USB wire from the serial USB port of the Raspberry Pi 3 B+ board to the Arduino Mega 2560 board. The wire is shown in Figure 5-28 for clarity.

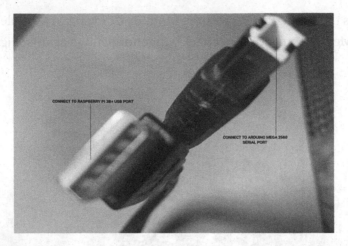

Figure 5-28. *USB serial cable to connect the Arduino Mega 2560 with the Raspberry Pi 3 B+*

Now you are finally ready to power on your complete system. The Arduino does not need a separate power cable, although it has a slot for it on the board. The serial cable powers on the Arduino Board because it is a slave of the Raspberry Pi 3 B+ board. When you power on the Raspberry Pi 3 B+, you should see LED lights on the Modbus Conversion Module and the Arduino Mega 2560 board. This signals preliminarily that your connections are successful. This can be seen in Figure 5-29.

Figure 5-29. *Final power on*

If the board or the module does not light up, you should check your connections again, especially the ones supplying power to the module or the Arduino board. Now go ahead and run the Arduino Hello world program from Chapter 3 to confirm the serial communication between the Arduino Mega slave and Raspberry Pi 3 B+ master.

Summary

This chapter was about preparing your setup for the case studies implementation. You are now fully configured by setting up the Raspbian OS on the Raspberry Pi 3 Model B+ and downloading it from the www. raspberrypi.org download options. You used the microSD card and microSD card adapter to get the OS. Then you used the Win32 Disk Manager to unpack and extract the package. After this, you used the HDMI-to-HDMI cable in order to connect the Raspberry Pi to a TV monitor. You used USB dongles to connect a keyboard and mouse to the SBC. After this, you connected the Raspberry Pi 3 B+ power cord to a power bank so that its power connection would be failproof in case of a power outage.

You then installed Python and tested it on Raspbian. Next you installed and verified the Thonny Python IDE. You then ran a Raspbian command line script to install the required Python packages

You ran a script to test the installation of the Python advanced packages and to check for errors. You installed the tkinter package on Python for a GUI implementation and then successfully executed tkinter on the Python 3.x interpreter. After this, you ran the iotpypackages.run Raspbian script to install the IoT Python packages on Raspbian.

After this, you finished installing the Arduino IDE by running the `install.sh` file. You installed a SQLite3 database for storing data from IoT sensors. After this, you installed the Modbus interface device, the single phase meter, onto the Arduino. You configured and connected the three devices, the Schneider Electric Conzerv EM6400 Series Power Meter, the 5V RS485 to TTL Signal Mutual Conversion Module for Arduino Overvoltage Protection with Signal Indicator, and the 5V Modbus Converter Module. After this, you made the Modbus converter wire connections to the Arduino Mega board and connections to the PWM and communications side of the Arduino Mega board. You wired the Modbus Conversion Module communication and PWM pins through the USB serial cable to connect the Arduino Mega 2560 with Raspberry Pi 3 B+. After this, you performed the final power on for your complete system.

With this you come to the end of setting up your system, including the hardware and software. You can now proceed with the case studies.

CHAPTER 6

Configuring the Energy Meter

In a continuation to Chapter 5 where you left off by connecting the three-phase Conzerv EM6400 Energy Meter from Schnieder Electric, in this chapter you will do two things. First, you'll write up the code for getting data from the EM6400 Energy Meter on the Arduino Mega slave and then reading it into Python on the Raspberry Pi master. The second is to set up another energy meter using single phase electrical connections and then read its data directly into Raspberry Pi via the USB serial Modbus protocol. The second setup is to enable those enthusiasts of the IIoT who want to get the feel of how real-life data is generated by industrial devices such as energy meters and machines. In the real world, however, you will see three-phase energy meters and machines like EM6400 communicating through the Modbus protocol. Again, I want to remind the readers that we are using the Arduino Mega 2560 because it is a truly industrial grade SBC that supports connecting multiple devices like energy meters or industrial machines connected to each other serially or in a daisy wheel electrical configuration. If you have just a single device, regardless of single phase, two phase, or three phase, you can connect it directly to a Raspberry Pi 3B+ like I show you in this chapter. So let's get started writing the code to get data from the EM6400 Energy Meter on the Arduino. The code for this is available in the file named modbus01_EM6400_Tested.ino code bundle that you can download from www.PMAuthor.com/raspbian/. Remember to open it up in the Arduino IDE and configure and test the Arduino IDE like I showed you in Chapter 3 before executing the code.

Coding for the EM6400 Energy Meter

Installing the Modbus libraries in the Arduino IDE is the first step before you start running any code. This is required so that you get some of the C language header files. ModbusMaster.h and SPI.h and other libraries are available to communicate with Modbus through the Arduino slave to your energy meter. You can open the libraries as shown in Figures 6-1 through 6-3.

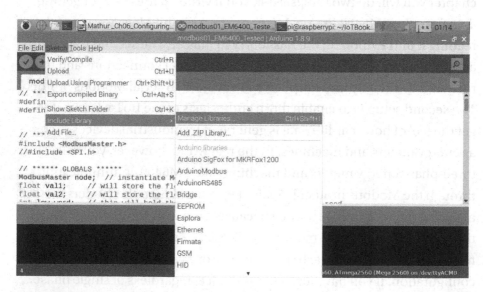

Figure 6-1. *Opening the Manage Libraries menu*

Open the Manage libraries menu from Menus Sketch ➤ Include Library ➤ Manage Libraries or use the keyboard shortcut of Control + Shift + I. The Library screen will open up, as shown in Figure 6-2.

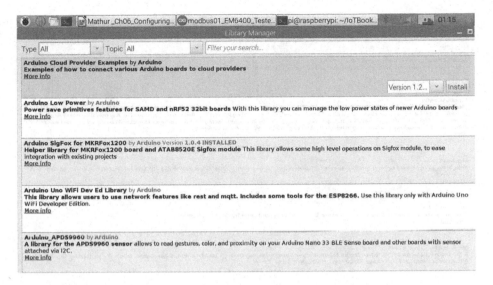

Figure 6-2. *Opening the Library Manage section in the Arduino IDE*

Now you have the library manager, which is the repository maintained inside the Arduino IDE for libraries installed in its environment. The wonderful things for a developer is that you do not need to worry about where you files are stored or kept or the paths associated with them. The library manager does the job itself; you just need to include the libraries exposed by it in your code. In your case, you need the Modbus library, which provide the two header files noted earlier, so your Arduino program can communicate with the energy meter. Type **Modbus** in the "Filter your search" field and press Enter to allow the library manager to get you all the libraries that have Modbus within their description. This is shown in Figure 6-3.

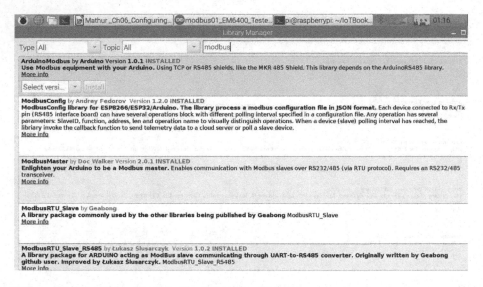

Figure 6-3. *Filter search for Modbus libraries*

As you can see in Figure 6-3, there's a listing of many libraries with Modbus in their name. You just need the first one, ArduinoModbus by Arduino, version 1.0.1. Note that you may see a different version when you do this in your IDE because the Arduino organization keeps updating its libraries by fixing bugs and offering enhancements to its users. After you have installed this Modbus library by clicking the Install button below its description, you will see a screen message saying "INSTALLED," as it says in Figure 6-3. Note that the "Install" button is greyed out because I already have it installed in my environment. After you have successfully done this, you can close the library manager window and return to your code window in the Arduino IDE, as shown in Figure 6-4.

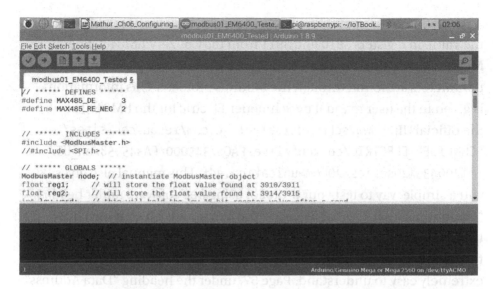

Figure 6-4. *Loading the code file named modbus01_EM6400_Tested.ino*

In the first few lines of code you see the include statement #include
<ModbusMaster.h>. This is the library that has now loaded. After the
include statements, you move ahead to declare the global variables, as
shown in Listing 6-1.

Listing 6-1. Code for Declaring Global Variables

```
// ****** GLOBALS ******
ModbusMaster node;   // instantiate ModbusMaster object
float reg1;          // will store the float value found at
                        3910/3911
float reg2;          // will store the float value found at
                        3914/3915
int low_word;        // this will hold the low 16-bit resgter
                        value after a read
int high_word;       // this will hold the high 16-bit resgter
                        value after a read
```

The first global variable is `reg1`, which is a float type variable to store the value in registers 3910 and 3911 from the Conzerv EM6400 Energy Meter. The registers are the actual places where values of various energy parameters are found through the Modbus protocol. This information is given in the user manual by Schnieder Electric for the EM6400 at the official URL: `www.schneider-electric.com/resources/sites/SCHNEIDER_ELECTRIC/content/live/FAQS/345000/FA345958/en_US/EM%206433%20series%20Communication.pdf`. This manual also gives you a simple way to test your protocol using free software under heading "Communication Test" on Page 55. You can try this on your own; it's beyond the scope of this book so I leave it up to you. The good thing is that the manual describes the entire process with screenshots and it's extremely easy to understand. Page 57, under the heading "Data Address" gives a detailed table with various electrical data parameters that the energy meter can transmit through Modbus protocol. This table has the individual parameter address and provides information on the register address number that you need to get that particular data. For example, for the data value of Line to Neutral Voltage value (VLN) the register address is 3911. The type of value is float so you use a corresponding variable in the code (Figure 6-5) as `float reg1` for register addresses 3910 and 3911 and `float reg2` for storing the values of registers 3914/3915. Please note each register is 16 bits so it will have a high word and low word each, so you declare the global variables `int low_word;` and `int high_word;` after it is read from the energy meter through the Arduino Mega 2560. Now that you understand the concept of registers and how data is transmitted back from the energy meter through the Arduino Mega, you can look at a union structure for converting them to float values. This is given in Listing 6-2.

Listing 6-2. Union Structure to Store Register Values from the Energy Meter

```
// converting to floating point - union structure
union u_tag {
  int bdata[2];
  float flotvalue;
} uniflot;
```

Here you declare an array of int variable bdata[2], which stores two values: the high and low words from a register. The float variable flotvalue is the one that actually converts and stores the values into a float data type because the actual value that it will be getting is a float data type, which is needed for the program to work. Usage of this union structure will become much clearer during the loop() function of the code, which is later. Now let's declare some functions that are going to help you complete the process of communicating with the energy meter using the Modbus interface protocol. The sequence of communication is given in Figure 6-5.

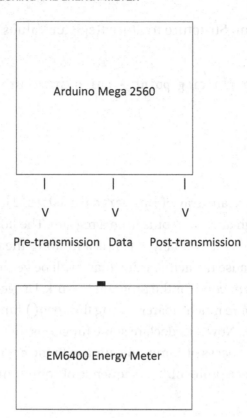

Figure 6-5. *Communication between the Arduino Mega 2560 and the energy meter*

As you can see, there are three transmissions that happen between the Arduino Mega and the energy meter. First there is a pre-transmission, then there is the actual data transmission, and then after the data is communicated back from the energy meter to the Arduino there is a post-transmission. This cycle is important to understand because the code is divided into these three sequences.

You must have noticed in Chapter 5 when you configured the Arduino Mega board and connected the pins to it that there were two pins, yellow and brown, that you connected to pins marked Rx and Tx. These are the ones where the actual communication happens. When data is read through these pins, there is a chip on the Arduino Mega 2560 board named ATMEL MEGA 16U2 whose LEDs light up; you can see them clearly communicating back to the Arduino board as a sign that your connections are correct. This, however, is not a guarantee that you will get back data; that depends on a number of factors such as the correct register and parameter numbers passed in the functions, etc. Pins 2 and 3 are connected to orange and green wires, respectively, on the TTL to RS485 converter module, which in turn is connected to the energy meter. It uses TTL communication to the energy meter; however, with a chip on its PCB it converts it back to the Modbus protocol and vice versa. To get data and to initialize them so that they get ready to transmit data to and fro, you need to define variables, as shown in Listing 6-3.

Listing 6-3. Defining the Pins for Communicating with the Energy Meter

```
// ****** DEFINES ******
#define MAX485_DE      3
#define MAX485_RE_NEG  2
```

Please remember that the DX or DE marked by the variable MAX485_DE is communicating through pin number 3 on the Arduino Mega board, which in turn is connected to the Modbus converter module on the corresponding DE male pin marked on it. Similarly, the variable MAX485_RE is negative and is communicating through pin 2 marked on the Arduino board, which in turn is connected to the Modbus converter module on the corresponding RE male pin marked on it.

During pre-transmission and post-transmission the main work of the next functions is to write to these pins to start and end communications. This is shown in the piece of code in Listing 6-4.

Listing 6-4. Pre- and Post-transmission Functions

```
// ****** Transmission Functions ******
void preTransmission()
{
  digitalWrite(MAX485_RE_NEG, 1);
  digitalWrite(MAX485_DE, 1);
  Serial.println("preTransmission");
}

void postTransmission()
{
  digitalWrite(MAX485_RE_NEG, 0);
  digitalWrite(MAX485_DE, 0);
  Serial.println("postTransmission");
}
```

The two functions given in the code initialize communication in the function void preTransmission(){} by using digitalWrite(MAX485_RE_NEG, 1);. This function initializes the RE pin# 2 to start writing. The first parameter in the digitalWrite function is the pin number, which is defined in the variable MAX485_RE_NEG, and the second parameter is the on/off binary parameter that enables writing to the pin with value of 1 and stops it by passing a value of 0. Next, MAX485_DE is initialized and then the postTransmission(){} function passes values of 0 to both pins to stop the transmission. Now that you have these functions, you are ready to execute the standard functions for the Arduino Mega, which are the void setup() and void loop(). To recall from Chapter 3, the void setup() function is defined first in the program in order to initialize and set up the code for the Arduino Mega 2560 to start functioning. For this function, you first initialize the output to pins 2 and 3 in order to start controlling the flow of data through them using the Modbus 485 protocol. Once this is done, you need to initialize the receiving mode for the same pins. After which you set the baud rate frequency at which the energy meter has been set to receive and send signals. The default for this meter is 9600 Hz and I don't intend to change it as it communicates well at this rate. After this, a slave id is defined for this slave. If you have multiple slaves connected to the Arduino Mega board, you can give multiple slave ids to them in this section. After this the preTransmission() and postTransmission() functions are called. As a measure to find out if the communication has happened correctly between the Arduino and energy meter, you can write something like "Hello World." This is exactly what you see in Listing 6-5.

Listing 6-5. The void setup() Function

```
// ****** STANDARD ARDUINO SETUP FUNCTION ******
void setup() {

  // make pins 2 and 3 output pins for Max485 flow control
  pinMode(MAX485_RE_NEG, OUTPUT);
  pinMode(MAX485_DE, OUTPUT);

  // Init in receive mode
  digitalWrite(MAX485_RE_NEG, 0);
  digitalWrite(MAX485_DE, 0);

  Serial.begin(9600);    // TX0/RX0 serial monitor
  Serial1.begin(9600);   // TX1/RX1 Modbus comms

  // Modbus slave ID = 1
  node.begin(1, Serial1);

  // Callbacks allow us to configure the RS485 transceiver
     correctly
  node.preTransmission(preTransmission);
  node.postTransmission(postTransmission);

  Serial.println("Hello World");

}
```

The Serial.println() function is sent as an output to the serial port by the Arduino. So you can use a Python program to read these statements from the Raspberry Pi master to which the Arduino Mega slave is connected. Now that you have your initial configuration for the Arduino done in the void setup() function, you can move on to the last step of defining the main work that you want the sketch to perform with the Arduino Mega. First, you need to read values from the energy meter and for that you need to define variables to hold the results and the data from it, as shown in Listing 6-6.

Listing 6-6. Code for void loop()

```
// ****** STANDARD ARDUINO LOOP FUNCTION ******
void loop() {

  uint8_t result;

  // Read Line to Neutral Voltage
  result = node.readHoldingRegisters(3910, 2);
   if (result == node.ku8MBSuccess)
  {

    high_word = node.getResponseBuffer(0x00);
    low_word = node.getResponseBuffer(0x01);

    uniflot.bdata[1] = low_word;    // Modbus data 16-bit low word
    uniflot.bdata[0] = high_word;   // Modbus data 16-bit high word

    reg1 = uniflot.flotvalue;

    Serial.print("Line to Neutral Voltage: ");
    Serial.println(reg1);

  }
```

```
//  node.clearResponseBuffer();
  delay(500);    // small delay between reads

  //Read Frequency
  result = node.readHoldingRegisters(3914, 2);
  if (result == node.ku8MBSuccess)
  {
    high_word = node.getResponseBuffer(0x00);
    low_word = node.getResponseBuffer(0x01);

    uniflot.bdata[1] = low_word;   // Modbus data 16-bit low word
    uniflot.bdata[0] = high_word;  // Modbus data 16-bit high word

    reg2 = uniflot.flotvalue;

    Serial.print("Frequency: ");
    Serial.println(reg2);
  }

  delay(5000);    // repeat reading every 5 seconds
  node.clearResponseBuffer();

}
```

Since this sketch is built on top of C and C++, the functions used in C and C++ and its datatypes also clearly integrate into this language. An unsigned integer of 8 bits is denoted by the datatype uint8_t in the sketch language, which is a compiler with wrapper on a C++ compiler. You define the result variable of this datatype to store the results of the holding registers, which you attempt to do in the next line. The first thing you read is the value of the neutral voltage from the holding register# 3910 using the statement result = node.readHoldingRegisters(3910, 2);. The node object, which is automatically made available in the sketch, is used to communicate with multiple Arduino slaves. In your case, there is only one and you have connected it to the Raspberry Pi. You initialize it using the statement node.begin(1, Serial1);, giving it a serial number slave id of 1 in the void setup() function. So now when you call the node object, the sketch automatically takes it to be this node that you initialized earlier. The second parameter value of 2 indicates the length in bytes of the register mentioned in the first parameter 3910.

The next piece of code uses an if statement to check if the response received from the Arduino slave was okay. In order to check this, you need to first know about the **Modbus function codes and exception codes** given in the Modbus master library. This is the Arduino library for communicating with Modbus slaves over RS232/485 (via the RTU protocol) (see http://4-20ma.io/ModbusMaster/group__constant.html). While this Arduino library has various variables to use, the specific one you want is ku8MBSuccess, which returns true only when the following checks are found to be okay in the communication between the Modbus master and slave:

1. The slave ID is valid.

2. The function code is valid.

3. The response code is valid.

4. Valid data is returned.

5. The CRC is valid. There is no error in transmission of data.

You can read more about this here: http://4-20ma.io/ModbusMaster/group__constant.html#ga81dd9e8d2936e369359777d67769a657.

Once this condition is tested as true by the if statement, you get the data values from the register by giving its starting and ending address in the statements high_word = node.getResponseBuffer(0x00); and low_word = node.getResponseBuffer(0x01);. For your reference, 0x00 is the hexadecimal value of 0 for high word and 0x01 is the hexadecimal value of 1 for the second low word. Now you need to concatenate the values of the high_word and low_word variables into a variable of type float. You create a union structure that has a float value variable by name of unifloat. The reg1 = uniflot.flotvalue; statement returns the complete float value of the first register and this is the value of Line to Neutral Voltage. You print out the value using the statement Serial.println(reg1);. Now that you have the value of Line to Neutral Voltage in a variable, you can get the other value of frequency, which is stored in register address 3914. You add a delay of 500 milliseconds; this is optional as it depends on what you wish to do with the data. If you are creating an alerting system that monitors the current throughput to the energy device, you would not want a delay since you want real-time data for both Line Neutral Voltage and the frequency of the current. However, polling the energy meter very frequently can create a load on the Raspberry Pi if you have multiple slaves, and this is an aspect of your application design you will need to keep in mind when you write this portion of the code. The code in Listing 6-7 gets the data from register address 3914, which is also 2 bytes long.

Listing 6-7. Code for Getting the Values of the Current Frequency

```
delay(500);   // small delay between reads

  //Read Frequency
  result = node.readHoldingRegisters(3914, 2);
  if (result == node.ku8MBSuccess)
  {
    high_word = node.getResponseBuffer(0x00);
    low_word = node.getResponseBuffer(0x01);

    uniflot.bdata[1] = low_word;   // Modbus data 16-bit low word
    uniflot.bdata[0] = high_word;  // Modbus data 16-bit high word
```

```
    reg2 = uniflot.flotvalue;

    Serial.print("Frequency: ");
    Serial.println(reg2);
}
```

You use the if condition to check if the data transmission from the energy meter to the Arduino slave is successful using the statement if (result == node.ku8MBSuccess). Once it is successful, the if statement block is very similar to the code block in Figure 6-11 except for the use of variable reg2, which you are using to store the float value of the second register, which contains the current frequency value inside the register. Using the statement Serial.println(reg2); you print out the value to the serial bus showing the frequency. Before closing the function void loop(), you add a delay of 2 seconds so that there is some time between successive reads. Again, this delay in time depends on the type of application you are building and can be changed as per the business need of the application. Just remember the less time delay that you have here, the more often system resources like CPU and memory your application are going to put on the three devices: Raspberry Pi 3 B+, Arduino Mega 2560, and the EM 6400 Energy Meter. After this, you clear the response buffer in the statement node.clearResponseBuffer();. This is to ensure that the old values are not collected again and new values come from the register to the buffer and then into the Arduino serial bus.

```
    delay(2000);      // repeat reading every 2 seconds
    node.clearResponseBuffer();
```

This concludes the code for programming the Arduino side of the program. You can compile it and upload it to the Arduino Mega 2560 by using the steps shown in Chapter 3 and Chapter 5. You should be able to see the code compile success and upload success messages like they appear in Figure 6-6 and 6-7.

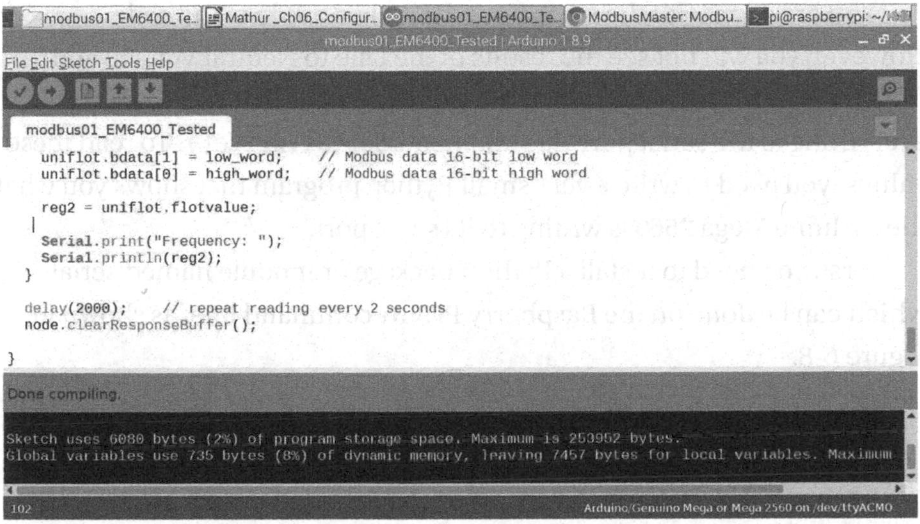

Figure 6-6. *"Done compiling" message after compiling the code for EM6400*

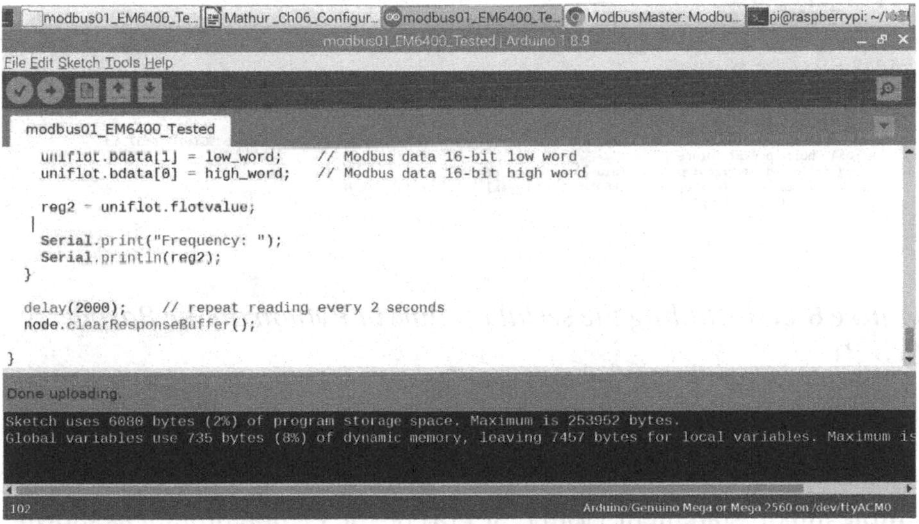

Figure 6-7. *"Done uploading" message after uploading the code for EM6400*

You are done with the steps from the Arduino side of the code. However, you will not see the results of the Line to Neutral Voltage and frequency in the Arduino side of things. Remember that you wrote everything to the serial port via statements `Serial.write()`. To read these values, you need to write a very small Python program that shows you what the Arduino Mega 2560 is writing to its serial port.

First, you need to install a Python package or module named serial, which can be done on the Raspberry Pi 3 B+ command line, as shown in Figure 6-8.

Figure 6-8. *Installing the serial module in Python on the Raspberry Pi 3 B+*

Python installations have this module by default, so you may not need to install it. You can check if it requires an installation by running the simple import statement `import serial` on the Python prompt as shown in Figure 6-9.

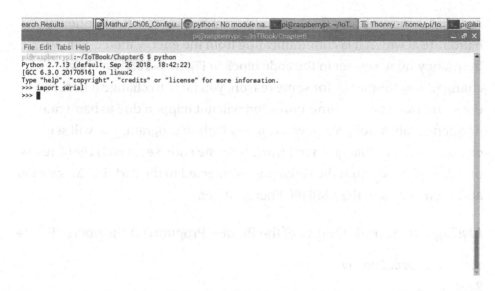

Figure 6-9. *Checking if the serial Python module is installed*

If you get an error, then the module for your version of Python is not installed. If you do not get an error and the Python prompt >>> returns without giving out any message, it's an indication that the library module is already present and you don't need to install it again.

Now that you have the serial library installed, you can go ahead and write the code for Python to fetch the data from serial port. See Listing 6-8.

Listing 6-8. Python Code to Read Data from the Arduino Serial Port

```
import serial
ser= serial.Serial('/dev/ttyACM0', 9600)
while 1:
    print("Reading...")
    print(ser.readline())
```

In this Python program, you first import the serial object and then create an object named ser with the serial.Serial() function by giving the first parameter, /dev/ttyACM0; usually this is the serial port where

Arduino communicates. The second parameter, 9600, is the frequency baud rate at which it is communicating from the energy meter. This is frequency value you set in the code block in Figure 6-10. If you have changed that frequency for some reason, you need to change it here to the same value or the communication will not happen due to baud rate frequency mismatch. When you run this Python program, you will see output similar to that given in Listing 6-9. The code `ser.readline()` reads one line of values from the serial port connected to the Arduino Mega 2560 and then in turn to the EM6400 Energy Meter.

Listing 6-9. Sample Output of the Python Program on Raspberry P 3 B+

```
>>> %Run arduino.py
Reading...
b'Hello World\r\n'
Reading...
b'preTransmission\r\n'
Reading...
Line to Neutral Voltage:
221.5
Frequency:
49.67
b'postTransmission\r\n'
```

You can see in the sample output that the program starts with preTransmission written and then prints "Line to Neutral Voltage:" and its value of 221.5 and then it prints "Frequency:" and its value of 49.67. This runs in an infinite loop every 2 seconds and every time the setup runs, "Hello World" is printed in the serial communication output. You can change this to something meaningful like "Setup initialized." I leave it up to you.

With this you come to the end of configuring and setting up the Conzerv EM6400 Energy Meter. Now, as I promised at the beginning of the chapter, you will get to configure an energy meter with just a single-phase two-wire setup, which is much simpler to set up and configure. This energy meter is by Eastron SDM630, which has a Modbus interface as well. This time you are not going to use the Arduino Mega 2560; you'll use Python and Raspberry Pi to directly get data from it.

Now you will first be connecting with a single Eastron SDM 630 through a single phase connection. I do not recommend connecting more than one of this meter either in a serial or daisy-wheel fashion because, unlike the Arduino Mega 2560, the Raspberry Pi 3 B+ is not a rugged industrial grade board and it can get burned easily. So the connection system that I am giving here is only for a small project, unlike the Conzerv EM6400 connection and configuration from earlier chapters.

The Eastron SDM630 Energy Meter (Figure 6-10) is made by a Chinese company and is easily available worldwide. You can check out the links on www.pmauthor.com/raspbian if you need one. This smart energy meter is one of the most popular ones for domestic and industrial purposes and supports three-phase connections as well. I recommend this one because I and my team have sufficiently tested it for personal and commercial applications for client projects and it performs well. Although there are a few competitors on the market, we use this popular one the most.

Figure 6-10. *Eastron SDM630*

The first step is to connect this energy meter to the Raspberry Pi and for that you will need the following components:

1. RS485 Modbus USB Convertor

2. At least 1 meter twin wire that can pass 5V current

3. Eastron SDM630 Energy Meter

4. Single phase machine such as a motor or a milling machine that has electronic coils and motors to produce reactive power

5. At least 20W LED bulb to produce active power

All the items listed are mentioned with their specifications at www.pmauthor.com/raspbian/. Let's get started connecting all of them together.

The RS485 USB Convertor needs to be connected with the twin wires, as shown in Figure 6-11. You can see the specific terminals to connect. There are two terminal pins at the base of the USB convertor.

Figure 6-11. *RS485 USB convertor*

They are marked as A and B. Please note that these are the same
terminals in the energy meter that are marked A and B, as shown in
Figure 6-12. If you insert the pink wire to the terminal marked A, then you
must connect the same pink wire at the other end to the energy meter
terminal marked A; if you do the opposite, it may damage the energy
meter, so keep this in mind. Similarly, connect the grey wire marked
at terminal B in the USB convertor to the one marked B in the energy
meter. This completes the USB convertor connection to the energy meter.
For your own safety, please ensure that neither the Raspberry Pi nor
the energy meter is running. Make sure they are in the off position and
unplugged from any electrical terminals. This is to prevent accidental
shocks. Also, never connect to the mains unless you have MCB or circuit
breakers installed in your electrical system, so that in the eventuality of
a short circuit the MCB trips and doesn't cause damage to the electrical
wiring system.

Figure 6-12. *Connecting wires to the RS485 Modbus Converter*

Have a look at the corresponding wire connections in Figure 6-13, which has the grey wire connected to the B- terminal and the pink wire connected to the A+ terminal on the Eastron SDM630 Energy Meter.

Figure 6-13. *Wires connected to the data terminals inside the Eastron SDM630*

Now that both the ends of the twin wires are connected, you must make two connections to the SDM630 Energy Meter. One is for the input wire connections, which will come from the mains and supply power to the energy meter, and the second is the output wire connections, which will supply electrical current to the load or the source of the current such as an LED bulb, electrical motor, or milling machine. This is depicted in Figure 6-14.

Figure 6-14. *Input connections to the Eastron SDM630 Energy Meter*

As you can see, there are four terminal wire connections in the input side of the meter. Out of the four terminals, since you are using a single-phase two-wire model in this energy meter, you need only terminal number 4 connected with the black wire or Neutral and terminal number 1 with the red wire or the power cable, also known as live wire. Once you have these wires connected, you can connect the socket end of these wires with a plug. Please note that you are not using the third wire, the ground wire, in this architecture. The plug connection is shown in Figure 6-15 for your convenience.

Figure 6-15. *Plug connection to the input wire of the energy meter*

You now have to make the connection to the output terminals of the energy meter and place a load at the other end of it like an LED bulb, electric motor, or milling machine. Remember they all need to be single phase. So connect them up as shown in Figure 6-16, where the output red or live wire is connected to terminal 6 and the output black or neutral wire is connected to terminal 8.

Figure 6-16. *Output wire connections to the energy meter*

Once this connection is done, you need to connect the other end of the terminal wire to a load such as an LED bulb, single-phase milling machine, or electric motor. In Figure 6-17, you can see an LED bulb connected as the load; however, this will not give very realistic data as the reactive power will always be zero. Nevertheless, it is a good way to test your circuit. The load at the other end can always be replaced with any other within no time.

Figure 6-17. *LED bulb connected to the output end of the energy meter*

At this stage, all the output and input connections are done and you can power on the energy meter by connecting it to the power socket from the input end. You should see the energy meter come to life and the LED bulb light up, as shown in Figures 6-18 and 6-19, respectively.

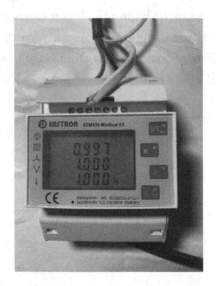

Figure 6-18. *Eastron SDM630 Energy Meter after power on*

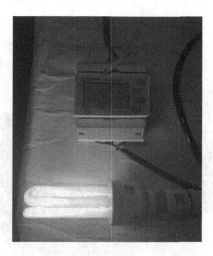

Figure 6-19. *LED bulb and energy meter after power on*

This completes the hardware side of the setup. Please remember that you have not yet connected the Raspberry Pi 3 B+ to the energy meter. This is just to check that all the connections to the energy meter are working and that the load powers on along with the energy meter. If this is successful, you can connect with the Raspberry Pi and then use a Python program to get data from it and store it in a SQLite3 database. After powering on the energy meter, you can power on the Raspberry Pi 3 B+. But before that, please remember to connect the RS485 Modbus converter to a USB port in the Raspberry Pi, as shown in Figure 6-20.

Figure 6-20. *Connecting the RS485 Modbus converter to the Raspberry Pi 3 B+*

Please note there is no LED light on this RS485 converter module PCB, so there is no way to tell that it is receiving a signal from the energy meter except from the Python software. It's named sdm630_tested.py and you can unzip it from the bundle pack available at www.pmauthor.com/raspbian/. The entire code is given in Listing 6-10. Before you can run the Python code, you need to install the Python libraries that are required by the Raspberry Pi to communicate with the energy meter using the RS485 Modbus protocol. For this you, need to install a Python library known as minimalmodbus. First, run the commands sudo apt-get update and then sudo apt-get upgrade because it is a good practice to keep the OS libraries and kernel updated and upgraded so that newer Python libraries do not fail due to any dependencies on the newer Raspbian structure. Figures 6-21 and 6-22 show this installation procedure.

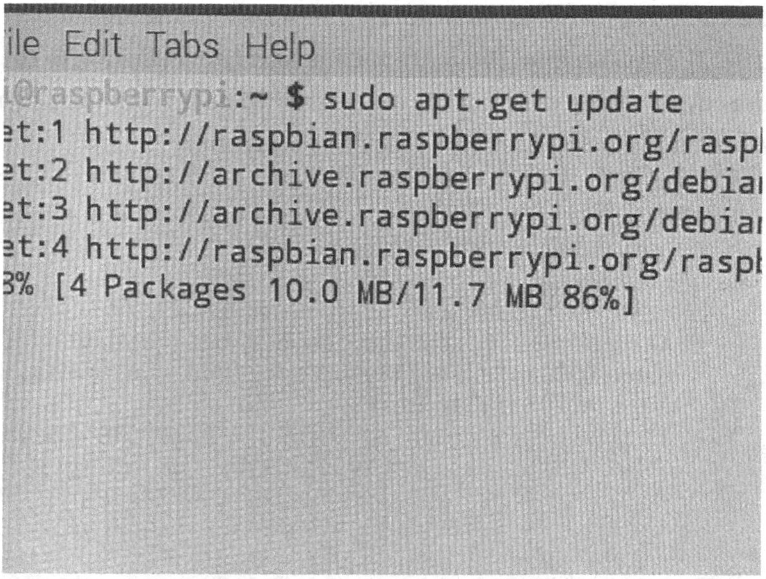

Figure 6-21. *Updating the Rapsbian kernel*

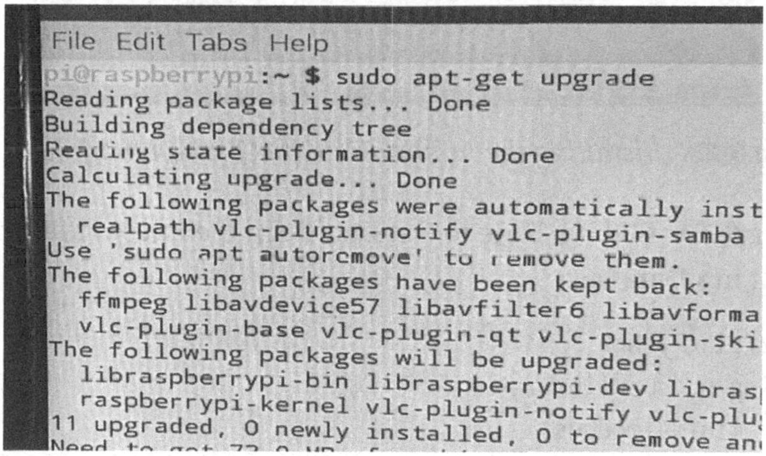

Figure 6-22. *Upgrading the Raspbian kernel*

After you have installed the minimalmodbus Python module
(Figure 6-23), you can load the Python code and run it (Listing 6-10). Just a
reminder: By this time your energy meter should be up and the Raspberry
Pi 3 B+ connected using the RS485 Modbus converter.

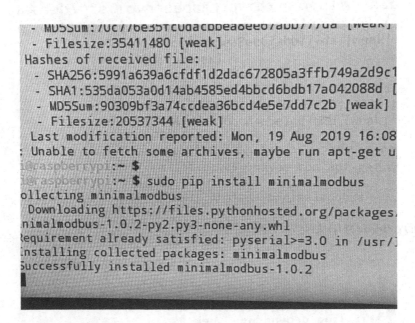

Figure 6-23. Installing the minimalmodbus Python module

Listing 6-10. Python Code to Get Data from the Energy Meter and
Store It in a Database

```
#!/usr/bin/Python
#Loading modbus library
import minimalmodbus
from datetime import datetime

#Initializing searial communication on the modbus library
sdm630 = minimalmodbus.Instrument('/dev/ttyUSB0', 1)
sdm630.serial.baudrate = 9600
sdm630.serial.bytesize = 8
```

```
sdm630.serial.parity = minimalmodbus.serial.PARITY_NONE
sdm630.serial.stopbits = 1
sdm630.serial.timeout = 1
sdm630.debug = False
sdm630.mode = minimalmodbus.MODE_RTU
print(sdm630)

while 1:
        Volts = sdm630.read_float(0, functioncode=4)
        Current = sdm630.read_float(6, functioncode=4)
        Active_Power = sdm630.read_float(12, functioncode=4)
        Apparent_Power = sdm630.read_float(18, functioncode=4)
        Reactive_Power = sdm630.read_float(24, functioncode=4)
        Power_Factor = sdm630.read_float(30, functioncode=4)
        Phase_Angle = sdm630.read_float(36, functioncode=4)
        Frequency = sdm630.read_float(70, functioncode=4)
        Import_Active_Energy = sdm630.read_float(72, functioncode=4)
        Export_Active_Energy = sdm630.read_float(74, functioncode=4)
        Import_Reactive_Energy = sdm630.read_float(76,
        functioncode=4)
        Export_Reactive_Energy = sdm630.read_float(78,
        functioncode=4)
        Total_Active_Energy = sdm630.read_float(342,
        functioncode=4)
        Total_Reactive_Energy = sdm630.read_float(344,
        functioncode=4)

        print('Voltage: {0:.1f} Volts'.format(Volts))
        print('Current: {0:.1f} Amps'.format(Current))
        print('Active power: {0:.1f} Watts'.format(Active_Power))
        print('Apparent power: {0:.1f} VoltAmps'.
        format(Apparent_Power))
```

```python
print('Reactive power: {0:.1f} VAr'.format(Reactive_
Power))
print('Power factor: {0:.1f}'.format(Power_Factor))
print('Phase angle: {0:.1f} Degree'.format(Phase_Angle))
print('Frequency: {0:.1f} Hz'.format(Frequency))
print('Import active energy: {0:.3f} Kwh'.
format(Import_Active_Energy))
print('Export active energy: {0:.3f} kwh'.
format(Export_Active_Energy))
print('Import reactive energy: {0:.3f} kvarh'.
format(Import_Reactive_Energy))
print('Export reactive energy: {0:.3f} kvarh'.
format(Export_Reactive_Energy))
print('Total active energy: {0:.3f} kwh'.format(Total_
Active_Energy))
print('Total reactive energy: {0:.3f} kvarh'.
format(Total_Reactive_Energy))
print('Current Yield (V*A): {0:.1f} Watt'.format(Volts
* Current))
import sqlite3
conn = sqlite3.connect('/home/pi/IoTBook/Chapter6/
energymeter.db')
#df.to_sql(name='tempdata', con=conn)
curr=conn.cursor()
query="INSERT INTO sdm630data(timestamp, voltage
, current ,activepow ,apparentpow ,reactivepow
,powerfactor ,phaseangle , frequency , impactiveng
, expactiveeng ,impreactiveeng , expreactiveeng
,totalactiveeng ,totalreactiveeng ,currentyield,
device ) VALUES(" + "'" + str(datetime.now()) +
"'" + "," + "'" + str(Volts) + "'" + "," +  "'" +
```

```
str(Current) + "'" + "," +  "'" + str(Active_Power) +
"'" + "," +  "'"  + str(Apparent_Power) + "'" + ","
+ "'" + str(Reactive_Power) +  "'" + "," + "'" +
str(Power_Factor) + "'" + "," + "'" + str(Phase_
Angle) + "'" + "," + "'" + str(Frequency) + "'"
+ "," + "'" + str(Import_Active_Energy) + "'"
+ "," + "'" + str(Export_Active_Energy) + "'" +
"," + "'" + str(Import_Reactive_Energy) + "'" +
"," + "'" + str(Export_Reactive_Energy) + "'" +
"," + "'" + str(Total_Active_Energy) + "'" + ","
+ "'" + str(Total_Reactive_Energy) + "'" + ","
+ "'" + str((Volts * Current)) + "'" + "," + "'" +
str("millmachine")+ "'" + ")"
print(query)
curr.execute(query)
conn.commit()
```

In this code, first you import the minimalmodbus Python library to
enable Modbus communication between the Raspberry Pi and the energy
meter. Then you import datetime to store a timestamp along with one
record of data from the energy meter. Its usage comes later in the code.
After this you initialize the Modbus instrument through its Rs485 Modbus
USB converter, which is connected on the Raspbian device list as /dev/tty/
USB0. This is the default device and should work in most cases; however,
if you face any communication errors, use the command lsusb to check
which tty port your 485 converter is connected to. Next, you initialize the
serial port communication at the frequency baud rate of 9600. This is the
default value; if you have changed the baud rate in your energy meter, you
need to change it here in the code statement sdm630.serial.baudrate
= 9600 to its new value. You initialize the byte size of 8 for a register
along with other setup parameters like stopbits and timeout interval. The
print(sdm630) prints the object out to the console on the Raspbian so

you should see all of the connection setup configuration through this print statement. After this, your setup of the energy meter is done and you can now communicate with it and get data from it. For this, in the next piece of code you use an infinite `while` loop to get the data continuously printing on the screen and then lastly store it in a SQLite 3 database. This code is very similar to the code for the Conzerv EM6400 Energy Meter so I won't delve deeper into it. However, the points to note are that you are getting a total of 15 values from the energy meter, including its voltage, current, active power, and reactive power to the current yield. After running the program, you should see output similar to Figure 6-24.

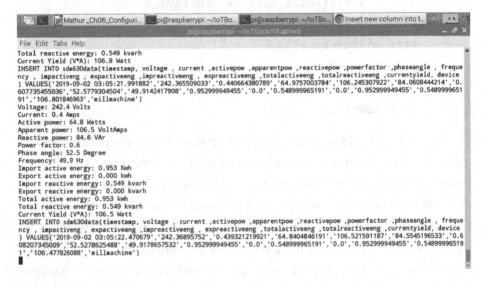

Figure 6-24. *Output of the program sdm630_tested.py*

The output of the `insert into` statement used to store data into the SQLite3 database goes into a table named `sdm630data`. The table creation script is in the file `sdm630_db_creation_script.sql`, which came along with the zip you downloaded from `www.pmauthor.com/raspbian/`.

Summary

This completes the setup, configuration, and getting data from the energy meter into a database system. So far you have successfully downloaded the scripts for use in the setup of the energy meter device. You coded for the EM6400 Energy Meter by the use of a Modbus library in the Arduino IDE. You also learned and implemented pre- and post-communication between the Arduino Mega 2560 and the Conzerv EM6400 Energy Meter by writing a small "Hello World" program. Then you wrote code to get data values from the energy meter through its registers. After this, you configured another energy meter, the more commonly available Eastron SDM630. You connected the hardware to it and then loaded the required Python libraries for the Modbus communication between the Raspberry Pi and the SDM360 Energy Meter. Lastly, you wrote a Python program to get data from the energy meter such as voltage, current, and power and stored it in a SQLite3 database.

Next, you will examine some case studies and design some interesting IoT and IIoT solutions.

Telecom Industry Case Study: Solving the Problem of Call Drops with the IoT

This case study is based on a fictitious nation state with a fictitious company and a fictitious scenario. However, what you will gain through this case study is a near-to-real problem of call drops, which most telecommunication companies grapple with. The case study takes you through the formation of a telecom operator, giving you enough background of its business establishment. The case study gives you a common political and business background and the socio-political stakes which are relatively uniform, so a specific country is not important for that. The reason for including the background of the political and business side in this case study is to make the machine learning engineer in you realize how important the decisions become based on the models for predictions that you build and how they can have a significant impact on the business and society at large. Let's get into the case study now.

© Puneet Mathur 2020
P. Mathur, *IoT Machine Learning Applications in Telecom, Energy, and Agriculture*,
https://doi.org/10.1007/978-1-4842-5549-0_7

Telecom Case Study Overview

Regoniatel started operations in 2016 as a 2G telecommunications operator in Regonia, a country surrounded by mountains on three sides.

The CEO is Kaun Methi John. He has other businesses, such as mining and refineries, through which he has become rich. He's established an empire in this small European country.

When Regonia adaptation the new 2G technology for mobile phone operations, the government called for proposals from interested businesses. John understood this was a golden opportunity for him to expand and get into this business of telecommunications, which was going to boom in the future. Since he did not have the technical expertise to set up a telecom infrastructure in his country, he actively looked for partners around the globe to help him achieve success in this new venture. He found Samtel, a USA-based company that was in setting up a joint venture with other companies.

Once the project plans and the feasibility study were carried out, the joint partners created a proposal for the 2G spectrum for the government. The bid was awarded to them and three other companies. They succeeded in getting the government contract; however, there was going to be stiff competition for setting up the telecom network and then getting customers onboard the network. The other three companies were also set up as joint ventures between local companies and international telecom companies. The government thought it wise to award contracts to such joint ventures, hoping the international experience could be taken advantage of.

The government carried out many months of testing of the 2G network in the country to determine its feasibility. Once this was done successfully by the internal telecom agency, the door was open for businesses to start their operations.

The booming 1 trillion economy of the country demanded an upgrade
in the communication infrastructure. The primary occupations in the
country were farming and mining. The country was surrounded by land on
all four sides; it did not have a seacoast. There was a lot of interest in the
gold and diamond mines this country was blessed to have, and they were
big revenue earners. The average income of a miner working in the gold or
diamond mines was $6,000 per annum. The reason for the higher wages
was that these mines produced some of the best gold and diamonds.
Diamonds from this country were in demand and sought the world over.
Mining contributed almost 50% to the GDP of Regonia.

The growing needs of this country required a good telecom
infrastructure to provide a robust communication system for its citizens.
This small nation could afford the best infrastructure in the world.

Regoniatel was able to set up good infrastructure in the urban areas in
the three years after it was awarded the spectrum contract. Its rivals, Ytel
and Fretel, also set up mobile towers in urban areas and rural areas. The
problem for Sautele after four years of developing an office infrastructure
was that it had become largely an urban-oriented company.

The government set up a telecom watchdog to publish data from
consumer interactions like customer complaints and network glitches for
telecom subscribers across all companies.

In its 2019 report, it highlighted a 35% jump in call drops across all
operators. The magnanimity of the problem reached such a level that the
ruling and opposition parties started mud-slinging matches on how the
telecom regulator and the government administration were shielding
the telecom operators. The Prime Minister called an urgent meeting with
the heads of the three companies in order to discuss the problem and its
possible solutions. Sautele received the following report about its call drop
complaints, which are given Table 7-1.

It clearly shows that has been a rise of call drop complaints on
Regoniatel's network by 41%. It had the largest spike in call drops across
all the operators. The telecom committee set up by the Prime Minister
suggested penalizing the telecom operators to the tune of $900 million
dollars for crossing a threshold of 10% complaints on call drops. This was
a heavy fine and would cost the company a huge amount of money if call
drops continued to happen on its network.

Looking at this threat and the public outcry, the president of
Regoniatel called for meeting with its operations team in order to
understand the key reasons behind the spike in call drops. Based on the
suggestions given by the operations manager, there was a requirement to
set up a pilot that would incorporate promising and innovative emerging
technologies like machine learning and the IoT in order to come out with
a solution to this techno-commercial problem. The operations manager
called in its most experienced machine learning engineer along with the
two best data scientists to work together on the pilot team in order to find a
solution.

The pilot team met for the first time and studied the various technical
reasons that were leading to call drops including studies on various
network locations stages given in Table 7-1.

Table 7-1. Call Drop Data for Telecom Providers of Regonia, Year 2018

Network Name	Average Signal Strength (dBM)	Subscribers	Teledensity %	Network Downtime %	Successful Calls %	Call Drop Rate	Calldrop Complaints
EtiTel	92	1160000	113	0.05	98.2	0.048	55680
GribaTel	107	1030900	96	0.87	97.4	0.8352	861007.68
RegoniaTel	89	1243080	97	1.3	96.2	1.248	1551363.84

As you can see, the signal strength in dB m which is the unit to measure signal strength. It tells about the number of subscribers teledensity percentage, network down time in percentage, successful calls in percentage the call drop rate also in percentage. The table also shows the number of call drop complaints received by each of the three telecom operators. The best average signal strength is that by GribaTel then by EtiTel and then followed by RegoniaTel. The highest number of subscribers are with RegoniaTel which are about 12.4 million. Teledensity is reported as 97% for RegoniaTel highest being with EtiTel. Highest network down time reported for RegoniaTel at 1.3%. The successful calls percentage highest for EtiTel and lowest with RegoniaTel. The call drop rate at 1.24 8% for RegoniaTel and lowest by EtiTel. Call drop complaint are also highest with RegoniaTel. So you clearly see a problem of network downtime, low average signal strength, and highest call drop rate with RegoniaTel.

The operations manager at RegoniaTel reported that the number of subscribers was going down and service customer feedback said that more than 60% describe switching to rival networks due to this call drop problem. The subscriber report is in Table 7-2.

Table 7-2. *Number of Subscribers of RegoniaTel for Seven Months*

Month	Number of Subscribers
Apr-18	1286588
May-18	1274157
Jun-18	1269806
Jul-18	1266580
Aug-18	1262845
Sep-18	1261726
Oct-18	1243080

As you can see, there is a spike from April 2018 to October and then a steady drop in subscribers. This is a steady drop of 3.4% of its subscriber base in a seven-month period. The president of the company expressed his concern to the operations manager and asked him to expedite the project in order to show some way out of this problem.

The machine learning engineer and the data scientists on this pilot were made aware of the seriousness of this problem and that it was directly linked to a reduction in revenue for RegoniaTel. If this exodus of subscribers continued, the subscriber base was projected to reduce by another 4.75% (Table 7-3). The operations manager also explained that every percentage drop in subscribers resulted in a corresponding drop of on average $10 million dollars in revenue. This was a substantial number for company that was struggling to keep the quality of its network intact. The CEO was already thinking that he might have to lay off some staff to reduce the operations cost. It was a scary thought but it was based on the numbers that were being presented to him. He explained to the operations manager and the pilot team how important the success of this pilot machine learning and IOT-based project was for the company.

Table 7-3. Projected Subscribers of RegoniaTel for 2019

Month	Number of Subscribers
Oct-18	1243080
Nov-18	1212003
Dec-18	1210760
Jan-19	1184034
Feb-19	1129338

The pilot team was asked to design and implement a pilot system
that would be able to address the direct problem of call drops and should
address the following concerns:

1. What was causing this problem of call drops?

2. In which location was this problem of call drops
    more prevalent?

3. How would you measure the call quality given a
    certain location?

4. Can a call drop be predicted? How?

This is the complete case study. If you are from the telecom industry,
you will find these issues very familiar and you should be able to relate to
them easily. Follow along with me to see how I would go about building a
solution as a machine learning consultant for this problem for RegoniaTel.

Setup and Solution for the Case Study

The setup tools are both hardware and software. The hardware elements
are the following:

1. An Android app to capture call data with a live
    Regoniatel SIM embedded on it

2. An industrial grade drone that could handle a
    mobile phone of 500 grams or 1 pound weight.

3. Raspberry Pi Model 3 B+ with a Wi-Fi module to
    receive data from the mounted mobile phone on the
    drone

The software elements:

1. An Android app that measures and captures call
   signal quality data and stores it in a CSV file format
   built by the team

2. A Python machine learning program with a
   prediction model from the imported data from the
   Android app device

Figure 7-1 shows how these elements come together to form the pilot
solution.

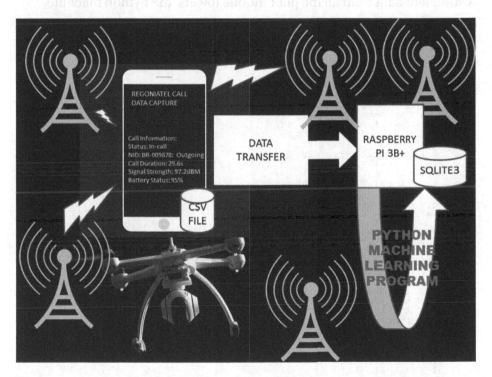

Figure 7-1. *Case study solution setup of hardware and software*

The key to this setup is the drone that carries the mobile phone around a particular area. The drone is manually piloted by a person on the ground and flown at human height level around the mobile towers where the calls are being piloted. The mounted mobile phone on the drone makes a call automatically every few minutes to another mobile phone in the same network. The purpose is to automatically capture the call quality data and store it into a data file in the CSV format. When the mobile phone has made a sufficient number of calls, it is brought down by the human operator and then taken to another location with a different mobile tower. This way the data is captured and then collated into a SQLite3 database. Once there is sufficient data from all the pilot mobile towers, the Python machine learning program is run to look at the results of the model. The mobile dataset is shown in Figure 7-2 and the Python is shown in Listing 7-1.

Call #	Netw ork1	CID 1	Netw ork2	CID 2	Date	Time	Speed	Outside Temperature	Outside Humidity	Signal Strength	Battery Level
	Rego	151	Rego	173	01/1	22:					
	niaT	306	niaT	784	0/20	14:	10				73
1	el	4	el	2	18	29	6	30	57	-60	.
	Rego	189	Rego	175	01/1	23:				-	
	niaT	180	niaT	933	0/20	38:				71.052	30
2	el	5	el	0	18	33	66	27	68	63158	
	Rego	172	Rego	176	01/1	07:				-	
	niaT	215	niaT	427	0/20	49:	11			81.578	45
3	el	6	el	1	18	06	3	31	78	94737	
	Rego	172	Rego	154	01/1	23:				-	
	niaT	590	niaT	590	0/20	25:				86.842	63
4	el	1	el	1	18	27	95	33	83	10526	
	Rego	142	Rego	157	01/1	02:				-	
	niaT	307	niaT	827	0/20	54:	10			92.105	59
5	el	7	el	5	18	43	5	35	88	26316	
	Rego	186	Rego	144	01/1	19:				-	
	niaT	957	niaT	059	0/20	50:				92.105	57
6	el	4	el	5	18	39	13	35	88	26316	
	Rego	175	Rego	177	01/1	07:				-	
	niaT	586	niaT	778	0/20	49:				81.578	43
7	el	6	el	8	18	46	2	31	78	94737	
	Rego	186	Rego	156	01/1	04:				-	
	niaT	350	niaT	336	0/20	34:					53
8	el	9	el	9	18	54	81	25	48	-50	
	Rego	197	Rego	183	01/1	21:				-	
	niaT	870	niaT	928	0/20	40:				89.473	39
9	el	8	el	1	18	34	92	34	85	68421	

Figure 7-2. *Mobile call quality dataset*

Please remember this is just a basic call quality dataset; in the real world, you would have a lot more parameters to be measured to determine call quality like the call duration, voice quality parameters, feedback over the phone, etc. The purpose of this dataset is to show you how to do an analysis from basic telecom data.

After importing this dataset into a SQLite3 database, we can use the program given in Listing 7-1.

Listing 7-1. Code for the Solution to the Case Study

```
# -*- coding: utf-8 -*-
"""
Author: Puneet Mathur
Copyright 2020
Free to copy this code with following attribution text: Author
Puneet Mathur, www.PMAuthor.com
"""

import pandas as pd
import sqlite3
conn = sqlite3.connect('C:\\ machinemon.db')
#df.to_sql(name='tempdata', con=conn)
curr=conn.cursor()
#query="INSERT INTO TEMPERATURE VALUES(" +"'" + str(datetime.
date(datetime.now())) + "'" +"," + "'" + str(datetime.
time(datetime.now())) + "'"+ "," + "'" + str(tem) + "'" + ","
+ "'" + tempstatus +"'" + ")"
df = pd.read_sql_query("select * from fielddata;", conn)
print(df)
#curr.execute(query)
#conn.commit()
```

```
#Looking at data
print(df.columns)
print(df.shape)
#Looking at datatypes
print(df.dtypes)
df.tail(1)

#Checking for missing values
print(df.isnull().any())

#EDA- Exploratory Data Analysis
import numpy as np
print("----------EDA STATISTICS----------------")
pd.option_context('display.max_columns', 40)
with pd.option_context('display.max_columns', 40):
    print(df.describe(include=[np.number]))

#Correlation results
print("----------Correlation----------------")
with pd.option_context('display.max_columns', 40):
    print(df.corr())

#Dividing data into features and target
target=df['Calldrop']
nm=['CID1','CID2','Speed','OutsideTemperature','OutsideHumidity',
'SignalStrength','BatteryLevel']
features=df[nm]
with pd.option_context('display.max_columns', 40):
    features.head(1)
    target.head(1)
```

```
#Building the Model
from sklearn.model_selection import train_test_split
x_train, x_test, y_train, y_test = train_test_split( features,
target, test_size=0.25, random_state=0)

from sklearn.linear_model import LogisticRegression

lr = LogisticRegression()
lr.fit(x_train, y_train)
# Returns a NumPy Array
# Predict for One Observation (image)
lr.predict(x_test)

predictions = lr.predict(x_test)

# Use score method to get accuracy of model
score = lr.score(x_test, y_test)
print(score)

import matplotlib.pyplot as plt
import seaborn as sns
from sklearn import metrics
import numpy as np

cm = metrics.confusion_matrix(y_test, predictions)
print(cm)

plt.figure(figsize=(9,9))
sns.heatmap(cm, annot=True, fmt=".3f", linewidths=.5,
square = True, cmap = 'Blues_r');
plt.ylabel('Actual Calldrops');
plt.xlabel('Predicted Calldrops');
all_sample_title = 'Accuracy Score: {0}'.format(score)
plt.title(all_sample_title, size = 15);
```

```python
plt.figure(figsize=(7,7))
plt.imshow(cm, interpolation='nearest', cmap='Pastel1')
plt.title('Confusion matrix', size = 15)
plt.colorbar()
tick_marks = np.arange(1)
plt.xticks(tick_marks, ["0", "1"], rotation=45, size = 15)
plt.yticks(tick_marks, ["0", "1"], size = 15)
plt.tight_layout()
plt.ylabel('Actual Calldrops', size = 15)
plt.xlabel('Predicted Calldrops', size = 15)
width, height = cm.shape
for x in range(width):
 for y in range(height):
  plt.annotate(str(cm[x][y]), xy=(y, x),
  horizontalalignment='center',
  verticalalignment='center')
plt.show()
```

Please remember that the code to get data from LightSensor and other IoT sensors is given in Figure 3-25 in Chapter 3. The process of getting data from the sensors remains the same regardless of the solution we are trying to implement. The only difference is that the Raspberry Pi in this case sits inside a drone while collecting data. It is the machine monitoring agent similar to Chapter 3's Figure 3-33 program to store LDR module IoT sensor data in a SQLite3 database and the Figure 3-35 code for simulating an IoT-based solution which together give us the ability to get data from IoT sensors and then store it and process it for advanced warning signals if required. In this case, we do not need any advanced warning signals.

The data is collected by the drone as it flies to the designated tower carrying the Raspberry Pi as its payload while the agent collects the data both from the IoT sensors for temperature and humidity and from the attached 4G/LTE module. After this data is collected and stored inside a

SQLite database on the Raspberry Pi, it is available for further analysis. So
this solution assumes data collection is already done using the setup.

The code starts with the usual import of pandas and SQLite3 libraries.
It builds a connection with the machinemon SQLite3 database and then
gets data from the database using the `Select * from fielddata` SQL
query. `Fielddata` is the table that stores data from the drone's IoT sensors
and 4G/LTE modules. Next, before building any module we check for
any missing values, which in our case are none so we do not need to do
any treatment for them. After this we do exploratory data analysis on the
dataset.

The problem with Python IDEs like Spyder is that they truncate
columns if we have a large number of them in our dataset. So the trick is
to use statement `pd.option_context('display.max_columns', 40)`
for the pandas dataframe. Here it's set to 40, but if you have larger number
of columns, you can set it to a value that you need. The output of the
`df.describe()` function is shown in Table 7-4.

Table 7-4. *Looking at EDA Statistics*

| | ----------EDA STATISTICS--------------- | | |
	Call#	CID1	CID2	Speed \
count	2000.000000	2.000000e+03	2.000000e+03	2000.000000
mean	1000.500000	1.693192e+06	1.688596e+06	59.706000
std	577.494589	1.709635e+05	1.673130e+05	35.214196
min	1.000000	1.399112e+06	1.398797e+06	0.000000
25%	500.750000	1.540734e+06	1.543660e+06	29.000000
50%	1000.500000	1.698208e+06	1.685488e+06	59.000000
75%	1500.250000	1.841730e+06	1.830084e+06	90.000000
max	2000.000000	1.987112e+06	1.986768e+06	120.000000

(continued)

Table 7-4. (*continued*)

	Outside Temperature	Outside Humidity	Signal Strength	BatteryLevel \
count	2000.000000	2000.000000	2000.000000	2000.000000
mean	28.046000	64.728000	-67.895895	52.089500
std	4.288482	12.714828	13.327106	19.043945
min	21.000000	40.000000	-92.105263	20.000000
25%	24.000000	55.000000	-78.947368	36.000000
50%	28.000000	63.000000	-66.000000	51.000000
75%	32.000000	75.000000	-57.894737	66.000000
max	35.000000	88.000000	-42.000000	96.000000

	Calldrop
count	2000.00000
mean	0.78900
std	0.40812
min	0.00000
25%	1.00000
50%	1.00000
75%	1.00000
max	1.00000

Some things to take note of are that the speed column values lie between a minimum speed of 0.0 in some areas to 120 in others; outside temperature lies between 21 degrees Celsius to maximum of 35 degrees Celsius; and outside humidity lies between 40% and 88%, which shows a fairly humid region. Signal strength ASU lies between -92.1 and -42.0; remember, the lower the better.

After completing this, we now look to see if there is any relationship between the columns statitically. The best way to do this is to run correlation over the numeric columns. df.corr()() is used here and it shows some good insights on what we can build our model on in Table 7-5.

Table 7-5. *Looking at Correlation*

```
n [102]: print("----------Correlation---------------")
with pd.option_context('display.max_columns', 40):
  print(df.corr())
----------Correlation---------------
```

	Call#	CID1	CID2	Speed \
Call#	1.000000	-0.028051	0.005757	-0.005541
CID1	-0.028051	1.000000	0.012414	-0.023611
CID2	0.005757	0.012414	1.000000	0.022053
Speed	-0.005541	-0.023611	0.022053	1.000000
Outside Temperature	-0.026658	0.013036	0.001161	-0.021969
OutsideHumidity	-0.016984	0.046798	0.018531	-0.025885
SignalStrength	0.016455	-0.047521	-0.018348	0.026063
BatteryLevel	0.017357	-0.023449	-0.017436	0.008438
Calldrop	-0.029110	0.038461	-0.001414	-0.045044

(*continued*)

Table 7-5. (*continued*)

	Outside Temperature	Outside Humidity	Signal Strength \
Call#	-0.026658	-0.016984	0.016455
CID1	0.013036	0.046798	-0.047521
CID2	0.001161	0.018531	-0.018348
Speed	-0.021969	-0.025885	0.026063
Outside Temperature	1.000000	0.771495	-0.772994
OutsideHumidity	0.771495	1.000000	-0.999792
SignalStrength	-0.772994	-0.999792	1.000000
BatteryLevel	0.032395	0.026747	-0.026404
Calldrop	0.496304	0.689299	-0.689721

	BatteryLevel	Calldrop
Call#	0.017357	-0.029110
CID1	-0.023449	0.038461
CID2	-0.017436	-0.001414
Speed	0.008438	-0.045044
Outside Temperature	0.032395	0.496304
OutsideHumidity	0.026747	0.689299
SignalStrength	-0.026404	-0.689721
BatteryLevel	1.000000	0.037316
Calldrop	0.037316	1.000000

There are quite a few things to note in this correlation. There is a significant positive correlation between outside temperature and humidity of 0.77 and a negative correlation between outside temperature and signal strength. This means when the temperature goes down, the signal strength also goes down (signal strength has an inverse effect so the lower the better). Call drop and outside temperature is 0.49, which is not very signifcant. The most signifcant thing to note is that there is a perfect negative correlation between humidity and signal strength. If the humidity goes up, there is definitely a drop in the signal strength. Practically, this means on a rainy day there would be low signal strength when the outside humidity is very high. Haven't we all experienced a drop in call voice quality during the rain? This dataset just proves it to be true.

The correlation between outside humidity and call drops is also positively high at 0.68, so this should also be considered for the model building. The correlation result to note for the model is between signal strength and call drops, which is a highly negative at -0.68. This means the better the signal strength, the less often the call drops. Now that we know our dataset holds potential, we can use these selected variables to make our machine learning model.

In the next step, we go about splitting our dataset into a target, which is the predictor and features the columns that are highly correlated to the target or predictor. For this purpose, we omit the following columns from our features: CID1, CID2, Speed, and BatteryLevel as they did not have a significant correlation in the results. We have selected only three variables for our model in the code: nm=['OutsideTemperature', 'OutsideHumidity', 'SignalStrength'], features=df[nm].

Further, we can start building our model by splitting the dataset into Training and Test through the x_train, x_test, y_train, y_test = train_test_split(features, target, test_size=0.25, random_state=0). I choose 25% as the test dataset and 75% as the training dataset. You can experiment further to see if you get a different result in your prediction model. Next, we use only the Logistic Regression classifier algorithm to build the prediction model;

221

however, I urge you to use a similar approach in selecting an appropriate classifier algorithm, as I did in my book *Machine Learning Applications using Python* in Chapter 3, Figure 3-32. Here we fit the dataset to our logistric regression object by using the code `lr.fit(x_train, y_train)` and then make predictions for the x_test, which is our features training dataset. We then score our predictions to see how accurately the model has performed, and the accuracy score for logsitic regression model is 1.0. This shows that our model has the problem of overfitting. So we need to evaluate other classifier algorithms, as I mentioned, to find out the best one that does not overfit.

The confusion matrix and the accuracy score can be visualized graphically by using a heatmap and colorbar. That is what we do next in the code by using the matplotlib and Seaborn libraries. The code builds a confusion matrix using a sklearn library metrics package through the statement `cm = metrics.confusion_matrix(y_test, predictions)`. Sometimes this simple output does not have an impact that a visually appealing color bar or heatmap can have as far as the confusion matrix is concerned and that is what is done in the next section of the code using the Seaborn heatmap library package in the code statement `sns.heatmap (cm, annot=True, fmt=".3f", linewidths=.5, square = True, cmap = 'Blues_r')`. Here the annotations are set to true so that the output can be seen on the heatmap; also we use `cmap` or a color map which has a blue hue scale. To learn more about annotated colormaps, refer to this this matplotlib URL: `https://matplotlib.org/gallery/images_contours_ and_fields/image_annotated_heatmap.html`.

Summary

In this chapter, you saw a near-practical real-world scenario of a fictitious telecom company going through a socio-political crisis and whose bosses were called upon to investigate and correct its problem of call drops.

Around the world telecommunication companies face similar technical
challenges and problems, and are using AI to solve them. I showed in
this case study solution how the problem of call drops can be predicted.
Predicting problems is the first hurdle that a business has to jump before
finding a solution for the problem. This was done in this case study. We
built a model using a logistic regression classifier algorithm to predict
when a call drop happens. Armed with this data, the business can see what
kind of technical upgrades are needed to address the humidity problem as
given by the prediction model. Are there better transmission and receiver
towers available that work well in such humid conditions? Do they need
to position some towers away from bodies of water? These are some of the
questions that the business will have to address once the prediction model
provides its results.

CHAPTER 8

Gantara power plant: Predictive Maintenance for an Industrial Machine

This chapter is about an energy industry case study where we try to simulate data for predictive maintenance for an industrial machine. The case study is against a backdrop of a fictitious country in the future in the African region. It lays down the challenges of setting up a power plant in a third-world country from taking up the task of land acquisition to conceptualizing its design to actually setting it up. Also covered are the technical requirements for setting up a power plant and the technical specifications, which include industry-relevant land for the power plant so that you understand what a power plant really needs into order to be set up. If you are new or have some exposure to the energy industry, what you will find helpful is the table with the input processing and output format of the requirements of a power plant, which lays down the complete structure and the base of the enterprise. Although the characters, the situation, and the country are all fictitious, they will give you an idea of how socio-political environment pressures work on a basic thing like electricity

© Puneet Mathur 2020
P. Mathur, *IoT Machine Learning Applications in Telecom, Energy, and Agriculture*,
https://doi.org/10.1007/978-1-4842-5549-0_8

energy generation. Also included as part of the case study is a machine
learning engineer whose character is carefully chiseled in the case study to
give you an understanding as to what pressures a typical machine learning
engineer would go through when working with these kinds of problems. So
let's dive into our case study.

The Case Study

Noki Ora, the CEO of a private power plant, has just landed in the capital
city of Gantara in Nobag, Africa, which is seeing rapid development due
to the progressive policies of its recently elected government. However,
with development comes the challenge of producing enough power for
industrial and residential needs. Noki has been with the company right
from the time the plant was conceptualized to its current state where it is
now generating power on peak loads of 500 MW. The journey has not been
straightforward; it's been riddled with problems all the way. The fact is
that power generation plant setup in a third-world country has immense
challenges from finding funding partners to finding plant equipment
suppliers willing to do the installation.

Nobag became an independent country in the African continent after
the end of the Syrian War in the year 2024 when the major countries came
together to carve out this nation. Ever since its independence, the country
has seen a rise in interest in the development of its mining and agrarian
industries, which are the mainstay occupations for the people here. Its
government has set up various special economic zones that give various tax
breaks to companies that come here to set up businesses. Noki's company,
Nobag Power Enterprise Ltd., is one such company that has been set up
by a group of European conglomerates in order to harness the country's
development potential and provide a boost to its infrastructure. Nobag

President Ag Zisi gave the following charter, which was part of the presidential
order, to Nobag Power Enterprises in order to succeed in its mission:

*Nobag Power Enterprises is given permission to generate and distribute
thermal power up to 500 MW capacity to support the capital region of
Gantara for its power requirements in a two year time period.*

The company's vision statement reflects the President's charter and
reads

*Nobag Power Enterprises has a vision to generate and distribute power
to meet the electrical energy needs of the people of Nobag in order to help its
residential and industrial development.*

Project Background

When the Gantara power plant was conceptualized by Nobag Power
Enterprises, it drew up the land requirement plan for construction and
installation of the power plant as this was the key factor for setting up
the project by the Nobag government. The requirement of land was
communicated to the government officials because this was a thermal
power plant and it needed the land acquisition to be completed quickly so
that it could start the power plant building process. The land plan schedule
for the 500MW power plant is given in Table 8-1.

Table 8-1. Power Plant Land Requirement for the Gantara Power Plant

Area in acres	
Description	**500MW**
Facilities inside the power plant boundary	
Main plant	25
Coal handling system	230
Water system	45
Water reservoir	30
Switchyard	25
BOP, stores, and road	70
Green belt area	150
Facilities outside the power plant boundary	
Ash disposal area	270
Employee township	100
Ash, raw water, and coal disposal	250
Grand total	1195

Although Nobag had ample land to set up a power plant project, most of the land around the city of Gantara was occupied by farmers. They gave stiff resistance to the newly formed Nobag government's efforts to acquire land for development of a power plant. This led to violence and the government declared an emergency in the area where the land acquisition was sought by the company as it was close to a water reservoir. Finally, after the government agreed to the farmers' demand for adequate compensation, the land acquisition went ahead after a struggle of nine months, which was a record of sorts given similar cases of government attempts to acquire land for dams and other projects; it had run into

litigation and violent protests. Although the people of Nobag understood
that the power plant was for their benefit, as was communicated by the
government and Noki Ora in his interactions with the farmers and the
people of Gantara, the issue was that of trust that the government was
giving them a win-win deal. With a bit of pressure from its European
conglomerate, the government bent to the pressure to acquire the land by
giving market-based compensation.

Noki Ora, 54, is an African-born electrical engineer who had worked
with companies like GE and had ample experience in setting up power
plant facilities for its European conglomerate partner. He had risen into
the CEO position after his successful implementation of a plant in Middle
East. He knew the geography well, being a native from this continent. He
was well connected politically and was able to pull a few strings in order to
make the project successful. He was shortlisted by the selection committee
among three potential candidates who had similar profiles but did not
have the ethnicity which gave him the advantage for this job.

Nobag has extensive coal mines reserves to last it another 100 years.
So setting up a thermal power plant is the best option as it has less
petroleum reserves to fulfil the requirement of natural gas for a power
plant. It will have to buy it from other Middle Eastern countries, which will
be expensive and will drain its already meager foreign exchange reserves.
Moreover, it will make it dependant on the natural gas supply for its power
because the neighbouring countries had just come out of a long war and
a natural gas pipeline could become the target of a militant strike if the
situation became fluid in the region again. Although Nobag has its own
pollution control laws, since it is taking help from external parties like the
European conglomerate, it must adhere to the clean power requirements
set by them. The thermal coal power plant must have CCS implemented,
which is the process of capturing the carbon and putting it into the ground
bed (https://en.wikipedia.org/wiki/Carbon_capture_and_storage),
thus reducing the release of carbon into the atmosphere. The primary
purpose of the power plant is to convert the chemical energy of the coal

into electrical energy. This is typically done by raising steam in the boilers, expanding it through the steam turbine, and then coupling the steam turbines to the electrical generators that convert mechanical energy into electrical energy. The input and output capacity of this thermal power plant is given in Table 8-2. Note that all figures are for a daily consumption cycle of 24 hours.

Table 8-2. Input and Output of the Gantara Power Plant

Description	Input	Processing	Description	Output
Coal	6000		CO2	15000
Furnace Oil	50.5	500 MW	SO2 + NO2	340
Water	4900		GT	460
Electricity	40		ASH	2100

The problem with a typical coal-based thermal power plant is that of generation of gases like Co2, SO2, and NO2, which heavily pollute the environment. So these gases need to be treated by capturing them and putting them in the ground, especially CO2. This increased the cost of the project and the requirement of funds. The power plant itself requires electricity for its own consumption and so takes in about 40 megawatts of current for its operation and only 460 megawatts is released for further transmission to the city of Gantara. The requirement of 4900 cubic meters per day is the reason why the specific land near a water reservoir was sought and also where the farmland was posing a problem for its setup. The ash generated is 2100 tonnes per day and the government invited private parties to build factories to create concrete products and wallboards from it so as not to pollute the land area around the water reservoir because it was also used by the farmers. When the power plant started near Gantara, there were hardly any factories for production of Portland Pozzolana Cement (PPC), which could recycle the ash generated so it was being released into dry landfills near the power plant. The environmental activists raised protests about creating such an environmental hazard for the farmers around the power plant; however, the government assured them that the facilities to completely use the fly ash output from the power plant would be ready in a few years' time, reducing the environmental hazard around the proposed power plant.

To build the thermal power plant near Gantara, the company took a loan from a global investor consortium of $600 million to cater to the increasing load of power consumption in the localities nearby.

On the commissioning of the new power plant, which has an installed peak load capacity of 500 MW against a current peak load demand of 640 MW, the problems that have plagued its operation are high power system losses, low plant efficiency, erratic power supply, electricity theft, blackouts, and shortages of funds for power plant maintenance. Overall, the country's generation plants were unable to meet system demand over the past decade. It is currently buying electricity from a neighbouring country to which it has to make payments in US Dollars, which is depleting its vital foreign exchange reserves. Whenever there is an outage in this power plant, backup power is taken from the neighbouring country to supply power during a blackout. It had issues with power generation during summer and winter peak load seasons.

President Ag Zisi of Nobag called for an urgent meeting last month with CEO Noki Ora based on heavy consumer complaints during the summer months about the power plants outages, which led to several hours of blackout in the capital city. The last outage lasted for more than 48 hours and the power was restored to the entire city only after the neighbouring country gave alternative power. The President wanted to why a new power plant that was commissioned only eight months ago was having problems. Thermal power plants have a life beyond 30 years. Due to the mounting pressure through the press on his government's inability to supply electricity to the nation's capital adequately, the President called the European conglomerate officials to constitute an investigation and

submit a report on the cause of the frequent outages and how this problem
could be resolved. The commission took more than 20 days to study the
plant infrastructure. It had among its team a machine learning engineer
named Carl, who was also roped in to help the team determine the real
reason of these outages based on data available from the power plant. Carl
used Modbus RS485 interface devices to get data from the key power plant
machines over a period of two weeks. His findings were as follows:

- Outage is the most common issue in the power plant
  due to its steam turbine.

- Finding a way to reduce the outage duration needs to
  happen.

- The number of outages needs to be reduced.

- A prediction model needs to be built around what is
  loading now and the projected loading across several
  hours for the power plant.

A second national committee was set up by the government of Nobag
based on the findings of the first commission to find out and rectify the
main reasons for the outages. The report found out that the fault was
actually due to lack of coordination on the operation of the gas turbine.

The following parameters were laid down by Carl, who was also part
of the second national commission, to collect outage data from the gas
turbine/grid:

- Total period, T

- Number of faults that occurred during period T

- Number of failures between maintenance

- Total operating time between maintenance

- Total outage hours per year

- Number of failures per year

- Uptime of gas turbine

- Downtime of gas turbine

As in any traditional coal-fired power plant facility, the devices and machines work in silos with no connected single source of data stored centrally. Each unit, from turbine to transformers to boiler, is monitored independently. Also the data asked for by Carl was not being stored by the power plant so it was not possible for the commission members to determine the real reason for the failure that occurred in the gas turbine. Noki Ora's company was asked by the national commission to collect the data using IoT sensors from the gas turbine and relay and store that data in new private cloud servers. The results were to be generated over a period of two weeks 24x7 so that the outage durations could be clearly recorded. The purpose was to help Carl create a model to predict an outage. Of course, describing and diagnosing an outage was part of the process. The effort by the commission was to make the power plant company become forward-looking rather than just backward-looking and to become predictive and prescriptive on the outages that were happening inside the power plant.

As with any new exercise to change the way operations worked, there was resistance to what was being done by the commission. The unions got together and were keen to know what the outcome of this exercise would be. Would it lead to job losses? Would it lead to cuts in pay? Everyone was curious. There was resistance to putting new IoT sensors around the turbines in order to get a new set of data. In spite of all this, Carl and his team were determined to go ahead and get their plan implemented.

The plan employed many sensors, such as the ones measuring temperature and light around the turbines to the ones that measured ultrasonic sounds near it. Carl wanted to ensure that he did not leave any stone unturned in determining the real cause for the turbine failures.

233

Carl, although a machine learning engineer, had to study the domain in which he was working, the power plant, so he had to learn the complete functioning of the power plant. He produced a short schematic of the power plant, and its units are shown in Figure 8-1.

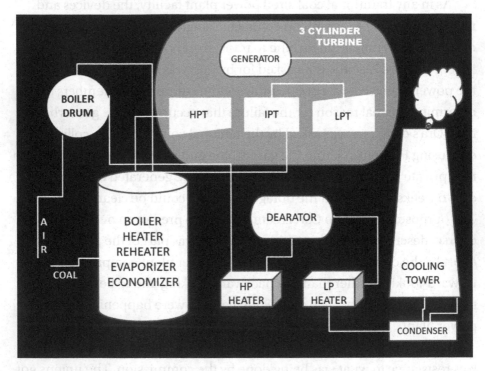

Figure 8-1. *Thermal power plant schematic diagram*

He showed this report to the commission officials and other members who were not aware of how a thermal power plan worked. His report read as follows:

"Please remember that this is not a technical diagram but a rough one to help you understand the key components of a thermal power plant. The actual power plant is more complex; however, the components remain the same. Let's discuss each of the components one by one.

A boiler is a vessel into which water is fed as an input and steam is taken out at desired pressure, temperature, and flow. Heat is applied on the container. For that, the boiler should have a facility to burn fuel and release the heat. The main functions of a boiler are as follows:

- Conversion of the chemical energy of the fuel such as air into heat energy.

- Conversion of this heat energy to water for evaporation as well to steam for superheating.

The basic components of a boiler are

- Furnace and burners

- Steam and superheating

Similarly, the other component, the economizer, is to improve the efficiency of the boiler by extracting heat from flue gases to heat water and send it to the boiler drum.

The advantages of the economizer, as the name says, include

- Fuel savings: Used to save fuel and increase the overall efficiency of boiler plant

- Reducing the size of the boiler: Since the feed water is preheated in the Economizer and then enters the boiler tube at a higher temperature, the heat transfer area required for evaporation is reduced considerably.

In the LP and HP heaters, the heat is taken out from the flue gases as they flow out of economizer and then further utilized for preheating the air before supplying it to the combustion chamber. It is key equipment for sending hot air for drying the coal in pulverized fuel systems to facilitate the combustion of fuel in the furnace. If this function is not present, there can be malfunction in the combustion chambers. Power plant furnaces also have a reheater component, which has tubes heated by hot flue gases flowing out of the tubes. Exhaust steam flowing out of the high pressure turbine is made to flow inside the reheater tubes to get energized in order to run the intermediate lower pressure turbines.

Steam turbines are used mainly as key components in all thermal power stations. The steam turbines are of two types:

- Impulse turbine

- Impulse-reaction turbine

The turbine generator consists of a series of steam turbines that are connected to each other and a generator on a common shaft (iron rod structure). There is a high pressure turbine (HPT) at one end, followed by an intermediate pressure turbine (IPT), two low pressure turbines (LPT), and the generator. The steam flows at high temperature in the range of (536 'c to 540 'c) and with pressure (140 to 170 kg/cm2) is expanded in the turbine.

The condenser condenses the steam flowing out of the exhaust of the turbine into liquid to allow it to be pumped further. The functions of a condenser are

- To provide the lowest rate of heat rejection temperature for steam.

- To convert the exhaust steam coming from the turbines to water for reserve, thus saving on the feed water requirement for the power plant.

The cooling tower is a semi-enclosed device for evaporative cooling of water by contact with air. The hot water coming out from the condenser is fed to the tower on the top and allowed to trickle in the form of thin sheets or drops. The air flows from bottom of the tower or perpendicular to the direction of water flow and then exhausts to the atmosphere after effective cooling.

The generator is the electrical end of a turbo-generator set. It is a convertor of the mechanical energy of the gas turbine into electricity. The generation of electricity is based on the principle of electromagnetic induction. Without the generator, the output of the gas turbine will not give out electricity. So it is the most important piece of equipment in the power plant."

After explaining this in his report, Carl explained the setup needed to get data from the IoT sensors around the gas turbine. The outage was happening in the gas turbine, as the first commission's report pointed out. However, it was not able to conclude the root cause or point to any potential problems that could have led to the repeated outage in the gas turbine. This made it more complicated, so it needed to be monitored along with the data that was being generated by the attached electronic unit with the gas turbine and was sent to a Raspberry Pi station to collect data using the Modbus protocol. In order to determine the root cause and to see if there was any effect of outside environment temperature, some IoT sensors were installed on the gas turbine. Three parameters were monitored: ambient light, ambient temperature, and an IoT sensor to measure ultrasonic sounds around the machine equipment. Of course, it was just a hypothesis at this stage that any of these elements from the outside environment could lead to an outage that was puzzling the plant engineers and the equipment manufacturers. Only meticulously collected data would be able to prove or disprove the hypothesis. The setup is shown in Figure 8-2.

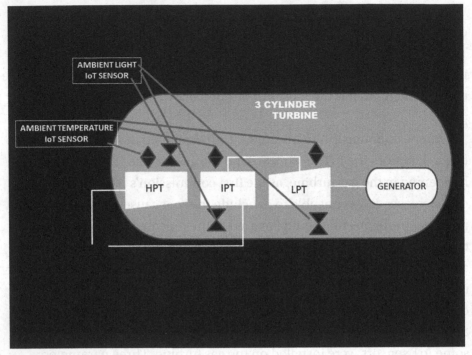

Figure 8-2. *Iot sensors used on the gas turbine to collect data*

The Raspberry Pi 3 B+ records the data using a Python program in the SQLite format for at least couple of weeks during which an outage in the gas turbine is likely to occur. The machine learning program is applied on this collected data using Python to arrive at final conclusions.

CEO Noki Ora wanted Carl and his team to provide the following results:

- Conclusive evidence of the cause of outage in the rogue steam turbine

- Possible solutions that would enable an automated process of monitoring and predicting outages

Both had to be demonstrated to the second commission as a proof of concept stage so that further commercial development could be taken up.

With this we come to the end of the case study. We will derive our
solutions from the dataset collected from the steam turbine machine in
the SQLite3 database named `powerplant.db`. We will be using this data
assuming that the power plant technicians collected this data from the IoT
sensors and sent it to the Raspberry Pi program. The entire solution with
Python code using machine learning is shown in Listing 8-1.

Listing 8-1. Code for the Solution to the Case Study

```
# -*- coding: utf-8 -*-
"""

Author: Puneet Mathur
Copyright 2020
Free to copy this code with following attribution text: Author
Puneet Mathur, www.PMAuthor.com
"""

import pandas as pd
import sqlite3
#Change the path as follows:
#On Raspbian: change to a path to your sqlite3 database file
with *.db extension
conn = sqlite3.connect('/home/pi/sqlite3/powerplant.db')
#df.to_sql(name='tempdata', con=conn)
curr=conn.cursor()
#The table with power plant data is named: TURBINE_INLET_
DATASET
df = pd.read_sql_query("select * from TURBINE_INLET_DATASET;",
conn)
print(df)
#curr.execute(query)
#conn.commit()
```

```python
#Looking at data
print(df.columns)
print(df.shape)
#Looking at datatypes
print(df.dtypes)
df.tail(1)

#Checking for missing values
print(df.isnull().any())

#EDA- Exploratory Data Analysis
import numpy as np
print("----------EDA STATISTICS---------------")
pd.option_context('display.max_columns', 40)
with pd.option_context('display.max_columns', 40):
    print(df.describe(include=[np.number]))
#Looking at the distribution graphically
df.hist(figsize=(10, 6))

#Correlation results
print("----------Correlation---------------")
with pd.option_context('display.max_columns', 40):
    print(df.corr())

#Dividing data into features and target
target=df['Defect']
nm=['Voltage','Current','Temperature','InletGap_mm']
features=df[nm]
with pd.option_context('display.max_columns', 40):
    features.head(1)
    target.head(1)
```

```python
#Building the Model
from sklearn.model_selection import train_test_split
x_train, x_test, y_train, y_test = train_test_split( features,
target, test_size=0.25, random_state=0)

from sklearn.linear_model import LogisticRegression

lr  = LogisticRegression()
lr.fit(x_train, y_train)
# Returns a NumPy Array
# Predict for One Observation (image)
lr.predict(x_test)

predictions = lr.predict(x_test)

# Use score method to get accuracy of model
score = lr.score(x_test, y_test)
print(score)

import matplotlib.pyplot as plt
import seaborn as sns
from sklearn import metrics
import numpy as np

cm = metrics.confusion_matrix(y_test, predictions)
print(cm)

plt.figure(figsize=(9,9))
sns.heatmap(cm, annot=True, fmt=".3f", linewidths=.5,
square = True, cmap = 'Blues_r');
plt.ylabel('Actual Defects');
plt.xlabel('Predicted Defects');
all_sample_title = 'Accuracy Score: {0}'.format(score)
plt.title(all_sample_title, size = 15);
```

```python
plt.figure(figsize=(7,7))
plt.imshow(cm, interpolation='nearest', cmap='Pastel1')
plt.title('Confusion matrix', size = 15)
plt.colorbar()
tick_marks = np.arange(2)
plt.xticks(tick_marks, ["0", "1"], rotation=45, size = 15)
plt.yticks(tick_marks, ["0", "1"], size = 15)
plt.tight_layout()
plt.ylabel('Actual Defects', size = 15)
plt.xlabel('Predicted Defects', size = 15)
width, height = cm.shape
for x in range(width):
 for y in range(height):
  plt.annotate(str(cm[x][y]), xy=(y, x),
  horizontalalignment='center',
  verticalalignment='center')
plt.show()
```

As you can see, the code implementation is exactly along the lines of the machine learning code implementation from in Chapter 3. However, do note that the results and the output are very different from the solution in Chapter 3.

Since we are replicating the machine code in Chapter 3, I will briefly describe the solution for this problem. An understanding of steam turbines is essential so let's look at them as part of this case study solution.

A thermal power plant has its prime mover medium as steam. To explain simply, water is heated, turned into steam, and spun in a steam turbine, which in turn drives an electrical generator. After this, the steam passes through the turbine and is condensed in a condenser; this is known as the Rankine cycle.

The steam turbine always operates under a high steam pressure
environment and has a number of high velocity moving parts. The nozzles
(inlet and outlet valves) and turbine blades are designed through a careful
analysis and its parts are manufactured to a high degree of precision.

Steam turbine plants generally have a history of achieving up to 95%
availability and can operate for more than a year between shutdowns for
maintenance and inspections. Their unplanned or forced outage rates
are typically less than 2%, or less than one week per year. In our case
study, after careful analysis of entire plant data, the problem of erractic
outages was narrowed down to a malfunctioning of a steam turbine by the
plant manufacturing engineers who were part of the group that included
machine learning engineer Carl.

In Listing 8-1, we first import libraries like pandas and sqlite3 for
connecting to the database. The database name is powerplant.db and
the table name is TURBINE_INLET_DATASET. We use this information to
connect to our power plant dataset by first building a connection to the
database using sqlite3.connect() and then reading all the rows from the
table TURBINE_INLET_DATASET using code pd.read_sql_query("select
* from TURBINE_INLET_DATASET;", conn). After we have successfully
read the data from the database to the pandas dataframe df, we then
have a look at its structure through the df.columns, df.shape, and df.
dtypes statements. After this, we look at the last row through the code df.
tail(1); we could have done a df.head(1) as well in order to see the first
row. This is just for a sanity check to see if the data has loaded properly.
The next check is to see if our dataset has any missing values, the chances
of which are less as the instrument used in our case to monitor the steam
turbine inlet and outlet valves is a Vibro Meter, which measures values
like the sound and the electrical inputs to the valve systems inside the
steam turbine. The code used is print(df.isnull().any()), which in
our case gives values of False as output since we do not have any empty

values in any of the columns. Once we pass this data check, we are good
to conduct exploratory data analysis on our power plant dataset. In the
practical world, you will need to merge data from the IoT sensors and
devices like the Vibro Meter and the steam turbine monitoring software
data such as GE's GateCycle™. This will be the most time consuming
activity for a machine learning engineer on the team. While merging you
may encounter data incompatibility issues such as different formats of
datetime, which you will need to address before the final merge happens.

In order to do a quick EDA, we use the `df.describe()` function of
the pandas dataframe. It gives us basic statistics such as mean, standard
deviation, variable minimum, variable maximum, data at 25th percentile,
50th percentile, and 75th percentile. After this, we look at the histogram of
the numeric columns, as shown in Figure 8-3.

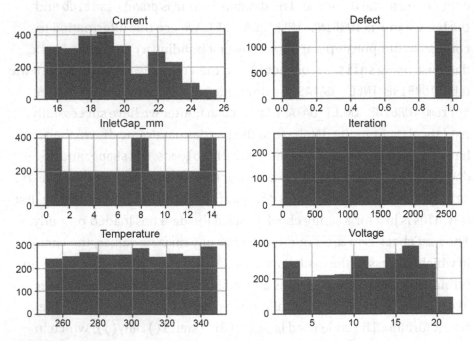

Figure 8-3. *Histogram of all the columns*

The Inlet Gap variable's curve is nearer to a normal curve; however,
it is heavy on the left side. Temperature is almost uniformly distributed.
However, the voltage variable shows some promise graphically. Let's
confirm the relationships through correlation. The statement to run is
df.corr() and the output is given in Table 8-3.

Table 8-3. *Result of Correlation in the Power Plant Dataset*

CORRELATION	Iteration	InletGap_mm	Voltage	Current	Temperature	Defect
Iteration	1	0.005	-0.1	-0.17	0.211	0.251
InletGap_mm	0.005	1	0.964	0.782	0.171	0.193
Voltage	-0.1	0.964	1	0.826	0.084	0.094
Current	-0.17	0.782	0.826	1	-0.176	-0.21
Temperature	0.211	0.171	0.084	-0.176	1	0.853
Defect	0.251	0.193	0.094	-0.21	0.853	1

We can infer the relationships that exist between the columns such as
Inlet gap and Voltage, which is highly correlated at 0.964; likewise columns
Inlet gap and Current at 0.782. Voltage and Current are also highly
correlated, as is expected, at 0.826. The inside temperature near the inlet
valve as measured through the industrial grade IoT temperature sensor
has a positive correlation with Defect at 0.853. The defects or anomaly
condition of the inlet valve operation of the steam turbine can be seen in
Figure 8-4 and the normal operations in the inlet valve operation cycle are
shown in Figure 8-5.

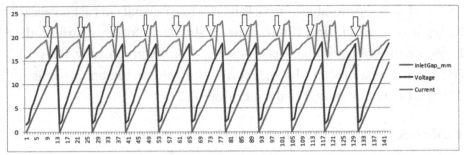

Figure 8-4. *Anomaly or defect in the inlet valve operation cycle*

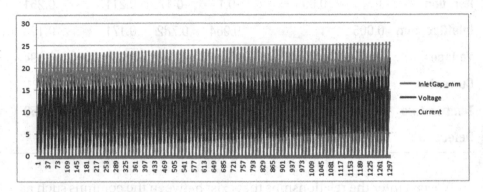

Figure 8-5. *Normal operations in the inlet valve operation cycle*

The red arrows on the inlet valve operation graph show the anomaly
in its operation between Inlet gap, Voltage, and Current. This shows the
conditions during which an outage in the power plant has occurred and
the root cause was fixed to a malfunctioning electronic controller to the
inlet valve of the steam turbine.

So what use is this data? Well, it allows a machine learning engineer to
build a prediction model for such a machine for not just anomaly detection
but also to give advanced warnings for predictive maintenance of the
steam turbine. Of course, this would require a lot more data than what you
saw here, but this is a good start for building a Proof of Concept that works
in a power plant operation. Let's continue building our prediction model.

In the next part of code, after calculating the correlations between the variables, we now move to dividing our dataset into target variable, the Defect column, and the features variables, namely Voltage, Temperature, and InletGap_mm. Note that I have not used cross validation (CV) or scaling as this is prototype-level code that just looks to see if building a model is feasible. You can add them as per Chapter 3 of my book *Machine Learning Applications Using Python*.

After this, we import the sklearn libraries like train_test_split and the classifier algorithms like Logistic Regression, Linear Discriminant Analysis, KNeighbors Classifier, Gaussian, Naïve Bayes, Decision Tree Classifier, and SVM. Each of the sklearn algorithm libraries are then initialized and then the fit() method is used for each of them to make the model learn using the x_train and y_train datasets. Once the system predicts, we are good to test its prediction accuracy on the test dataset. This is done using the predict() function of each of the classifiers. Next, using the accuracy_score() function, we score each of the algorithms and check their results. Please note that in the real world you will not get such an accurate prediction of 1.0 accuracy score for all the classifiers. So based on these results, you will need to choose the best classifier that gives most accurate results in the shortest time. Again, I discussed this in code format in my book *Machine Learning Applications using Python*.

Classifiers have categorical variables as predictors so they should have a confusion matrix and that is what we create both in text and graphical format as last part of the code to look at the results.

Summary

In this chapter, you have looked at a very technical case study belonging to the energy sector: that of a thermal power plant. This case study brought you to the forefront of some of the challenges faced by a power plant corporation in not just setting up a power plant but also running it successfully. This case study brought forward the problem of outages at a very young power plant, which most likely were due to temperature, as is the case with power plants in hot climates. The team used data to pinpoint the problem to a rogue inlet operation unit inside a steam turbine but needed data and a prediction model to create an advanced warning system for such anomalies. This would trigger the request for predictive maintenance whenever symptoms of such anomalies were seen through regular monitoring of data. We created a solution as a proof of concept if building such a model was feasible or not and also if there could be any tangible result. By using classification algorithms, we created a model that predicts 100% of the time or has an accuracy score of 1.0, which is rare in the real world. You saw how the entire solution can be created with ease with the use of data from IoT sensors and the Raspberry Pi Model 3 B+.

CHAPTER 9

Agriculture Industry Case Study: Predicting a Cash Crop Yield

This chapter covers an agriculture industry-based case study for predicting a cash crop yield. The case study gives you an idea of the challenges faced by a mid-sized agri-conglomerate trying to reach the next level and become as big as the vision of its founder. The problem of crop yield is very important for such organizations due to the fact that they want to maximize their land resources to get the highest revenue possible. What you will primarily learn in the case study is the fact that a company's vision should be tied to the machine learning operation that it is undertaking; otherwise it will be a wasteful expenditure. This is what I tell most of my clients when they hire me for consultation to look at how their current machine learning application helps them. Does it lower costs? If so, by how much to increase revenue? Then by how much? The goal of the project has to be quantifiable or it will not be successful but it will give dissatisfaction to the business owners and stakeholders. So read on...

© Puneet Mathur 2020
P. Mathur, *IoT Machine Learning Applications in Telecom, Energy, and Agriculture*,
https://doi.org/10.1007/978-1-4842-5549-0_9

Agriculture Industry Case Study Overview

An international agriculture conglomerate named Aystsaga Agro has investments in farming, fertilizer production, and tractors. The president, Tamio Polskab, is also the founder of the company. He has steered the company from its humble origins from one North American farm inherited from his father to an agri-business that spans three continents, North America, Africa, and Asia. The agri-business is not an easy one as it has many challenges that are emerging as a result of tariff wars and climate change. Worldwide, there is an agriculture crisis happening from South Asia to African countries trying to protect small farmers from the onslaught of such rapid changes. There is one big silent change that is happening: the use of technology, such as robotic machines, to replace humans in farming. Agriculture conglomerates such as Aystsaga Agro, which are highly commercial in their operations, are embracing technology to see how they can adopt and benefit from it. Since agriculture is not such a profitable business as compared to other industries such as service industries, the funding for such operations is really hard to get due to low ROI. This is the reason why such commercial companies focus on cash crops like corn, sugarcane, wheat, and soybeans. Sugarcane, sugar beets, and tomatoes fetch better yield than other crops for Aystsaga Agro, as shown in Table 9-1.

Table 9-1. *Global Yearly Production at Aystasaga Agro for Three Cash Crops*

Aystsaga Agro 2017-2018 Worldwide Production	Tonnes	Total Land Size
Sugarcane	470817.99	5361.70
Sugar beet	503999.80	5361.70
Tomatoes	273446.70	5361.70

In 2013, Aystsaga Agro purchased big farms in Brazil and South Africa. The Brazil farms were in the region of Sao Palo, and the South African farms were in the interior regions. Tamio, the founder of the company, succeeded in expanding the company's agriculture farming operations by acquiring land and using it for commercial operations. His vision was to establish his company as a major agri-business in the world.

While drinking his morning green tea, Tamio looked at the yearly financial results his CFO sent him that morning for his review. See Table 9-2.

Table 9-2. *Worldwide Production Numbers and Revenue Figures for Aystsaga Agro*

2017-2018 Worldwide Production	Tonnes	Total Land Size	US$
Sugarcane	470817.99	5361.70	241058809.3
Sugar beet	503999.80	5361.70	25199990
Tomatoes	178446.70	5361.70	180231167
Income from Agricultural Operation			446489966.3

While he was looking at the numbers, there was a knock on his plush South African office cabin door. He raised his head to catch his Chief Operations Officer, Glanzo, smiling. He signaled for him to come and sit in front of him. As Glanzo came and sat down, Tanio rose up and went near the right side of the window. He looked at the breathtaking view of the sea in the South African capital as he started speaking slowly.

The Problem

"Have you seen the numbers sent by Nambi this morning?" asked Tamio.

Glanzo nodded and said, "I think they are pretty impressive given that we have had several storms around our farms in Brazil and South Africa this year."

"You don't understand my vision, do you?" asked Tamio, looking straight at his right-hand man. "My vision is to achieve a turnover of USD 500 million and reach USD 1 billion in six years. I told the board about this in the last AGM. You were there too," said the boss.

Glanzo said, raising his shoulders, "We were all there, but I thought you were saying that to please the investors and the board."

"No, that was a goal that I have been working towards for the past 15 years, ever since this company went public. This is not just a vision or a goal; this is a challenge that I have thrown up to you all," said Tamio decidedly.

Glanzo retorted, "Yes, I understand the need to grow; otherwise, we will be eaten by the big fish in the industry. But you must understand the challenges that our operations are facing today, and unless we find real solutions for them, we will not grow at the rate that you want it to happen."

Tamio now had a frown at his face as he sat down and rested his head on his huge, leather executive chair. He was listening intently as Glanzo spoke further. "The single thing that is preventing us from growing is our ability to predict the yield of crops in a given land." Glanzo continued, "Our agriculture operations are a major problem for us in regions where we made a bad decision in buying infertile land that is low yielding. Low yielding land requires us to rectify it by applying different chemicals on hectares of land, which increases the cost of production and eats into our profitability. If we have to grow at the rate that you are spelling out, we need a way to determine which land is high yielding for a particular crop. If we are able to achieve this, we can avoid buying farms that are unproductive and low yielding, like the ones we have in Brazil where the soil is highly acidic and has to be treated with lime to make it more alkaline."

Tamio raised his head slightly, signaling that he understood what Glanzo was talking about. He asked, "Do you know of a way to find out how to predict the yield for a particular type of land, Glanzo?"

Glanzo responded, "I have been looking at some of the research that has been happening in universities around the world; however, nothing concrete is available. But, in my opinion, data machine learning and AI show some promise to solve the problem. We can hire some machine learning engineers and data scientists who can help us create a model for ascertaining the yield of crop in a particular land location."

"That sounds great. Why don't you form a team to look into this problem and propose possible solutions?" asked Tamio, smiling at Glanzo.

Glanzo said, "Yes, that is what I intend to do. First, I'll hire machine learning engineers and data scientist and then I'll add people from our business operations to the team."

"Send me an email for approval to go ahead," said Tamio, picking up the eyeglasses from his desk as Glanzo rose to leave his cabin.

Machine Learning to the Rescue?

The machine learning team was formed in two months' time with Hert Liu hired as the machine learning engineer for the pilot project. Along with three data scientists, he was made to take a robust tour of Aystsaga Agro's agriculture operations in Brazil, South Africa, and India. The business operations team members were introduced to them once their induction program was over. With detailed briefings, Hert and his team had various new words typical to agriculture farming added to their vocabulary. They also understood the inside processes that went into producing crops. The detailed tours really helped the team in dig deeper into the company's operations. However, what they lacked was the experiential knowledge, and that is the gap the business operations team members were going to fill in this pilot team.

Hert and his team met a couple of times after coming together. They met Glanzo several times too. He effectively communicated the company founder's vision and the problem at hand, which was linked to its growth. Hert was an experienced machine learning engineer who worked in the insurance domain earlier. His only brush with agriculture was creating a model for predicting claims for agricultural farmers. So this was a very high learning curve for him, where he had to understand the intricacies of the commercial farming business and also the intricacies involved in managing it, such as crop failures due to insect infestations, changes in weather patterns, and the soil constitution and its effects on farm output. Hert and his team started gathering some parameters that they felt could help in predicting the farm output of a crop. They divided their analysis into weather, soil, and economic environments. They shared their understanding with Glanzo in bi-weekly meeting with him. Hert showed how weather was damaging crops in their Indian farms due to unpredictable rains and floods near the farm. Glanzo pointed out that the biggest problem they faced was determining the profile of high yielding farm land before purchase. "As per our founder's vision, we are looking to buy a lot of farm land around the world such as in Australia, Thailand, and other regions; however, we can't do this simply by blindly buying land and then finding out it yields less crop that the average farming operations. This means disaster to our ROI. You as a team should build a model that helps us in determining the profile of high yield cash crop farmland. To do so, you should look at the soil profile and what makes soil give high yield or low yield. Use whatever instruments you want; we can buy them. There is no shortage of funds but you must build a system that will benefit us the most in our bid to grow exponentially," he said.

After getting clear direction on which way to proceed, Hert and his team got together with the business operations team members to understand what constituted a soil profile for a sugarcane crop, which they chose for their pilot project. They selected the parameters shown in Table 9-3.

Table 9-3. *Parameters for*
Monitoring Soil Nutrients

Soil pH
Organic Carbon %
Nitrogen kg/ha
Phosphorus kg/ha
Potassium kg/ha
Zinc mg/kg
Iron mg/kg
Copper mg/kg
Manganese mg/kg
Sulphur kg/ha

After having decided to zero in on these parameters in order to create a soil profile, they now wanted to collect data from each of the farms from Aystsaga Agro globally. Glanzo helped them buy the following equipment:

1. Commercial grade IoT sensors kits to read soil nutrient data

2. Soil nutrient manual testing kits for places where IoT sensors were hard to run

We are now going to build a solution for the problem of predicting the yield of a sugarcane cash crop based on the dataset from the file cashcrop_Yield_dataset.csv.

Solution

We can assume that the dataset is produced after reading soil samples from the IoT sensors and manual soil nutrient test kits for the various parameters given in the dataset file cashcrop_Yield_dataset.csv. The code in Listing 9-1 is not the most definitive solution for this problem but one that is simple and quick to achieve, as in any pilot project. In a real-world scenario, the data collected for soil nutrients will have many more parameters. Such a data collection exercise may take months to complete if the size of operations spans several continents. Agriculture operations are spread far and wide away from urban places so it's a hectic job for any company to comply. We'll keep these factors in mind when designing a solution. Note that we're only using linear regression as it gives a highly accurate score; however, I leave it up to you to try other regressor algorithms.

The Python Code

Listing 9-1 gives the Python-based solution for the problem that Aystsaga Agro is facing. The code starts with the usual imports of the common Python libraries such as pandas, StringIO, requests, etc. After this, we load the dataset from the CSV file cashcrop_Yield_dataset.csv. You can use a SQLite database instead. After this, we do the exploratory data analysis by looking at the mean, median, mode, standard deviation, and minimum and maximum values of each column of the dataframe. After this, we look at the outliers and count them, such as in the code df['Yield_per_ha']. loc[df['Yield_per_ha'] <=151.50000].count() for the yield per hectare column. This is going to be our predictor because the company wants to know the parameters that increase this yield, as discussed in the case study. We then visualize the data using boxplots and look at the skewness and kurtosis of the numeric columns. We also visualize it using an area plot and histogram. After this, it's time to look at the relationships between the columns with reference to the predictor column Yield_per_ha with the df.corr() code. Next in a three-step process, the code in the first step splits data into features and target variables. In step 2, it shuffles and splits the final dataset into training and testing datasets for building the prediction model. The last step is model building and evaluation, where we use linear regression since our goal is to predict a numerical variable yield per hectare. The last lines of the code give us the ability to take any new farm values and predict yield per hectare through the code predicted= regr.predict([[26,1500,6.8,0.9,367,32,490,35]]). To understand more about the result of the code and how it is being executed, look at the discussion after Listing 9-1.

Listing 9-1. Code for the Solution of the Case Study

```
# -*- coding: utf-8 -*-
"""
Created on Tue Oct 08 19:33:25 2019

@author: PUNEETMATHUR
"""

#importing python libraries
import pandas as pd
from io import StringIO
import requests
import os
os.getcwd()

#Loading dataset
```

```python
fname="cashcrop_Yield_dataset.csv"
agriculture= pd.read_csv(fname, low_memory=False, index_
col=False)
df= pd.DataFrame(agriculture)
print(df.head(1))

#Checking data sanctity
print(df.size)
print(df.shape)
print(df.columns)
df.dtypes

#Check if there are any columns with empty/null dataset
df.isnull().any()
#Checking how many columns have null values
df.info()

#Using individual functions to do EDA
#Checking out Statistical data Mean Median Mode correlation
df.mean()
df.median()
df.mode()

#How is the data distributed and detecting Outliers
df.std()
df.max()
df.min()
df.quantile(0.25)*1.5
df.quantile(0.75)*1.5

#How many Outliers in the Total Food ordered column
df.columns
df.dtypes
df.set_index(['FarmID'])
```

```python
df['Yield_per_ha'].loc[df['Yield_per_ha'] <=151.50000].count()
df['Yield_per_ha'].loc[df['Yield_per_ha'] >=159.285].count()

#Visualizing the dataset
df.boxplot(figsize=(17, 10))
df.plot.box(vert=False)
df.kurtosis()
df.skew()
import scipy.stats as sp
sp.skew(df['Yield_per_ha'])

#Visualizing dataset
df.plot()
df.hist(figsize=(10, 6))
df.plot.area()
df.plot.area(stacked=False)

#Now look at correlation and patterns
df.corr()

#Change to dataset columns and look at scatter plots closely
df.plot.scatter(x='Yield_per_ha', y='Soil_pH',s=df['Yield_per_
ha']*2)
df.plot.hexbin(x='Yield_per_ha', y='Soil_pH', gridsize=20)

#Data Preparation Steps
#Step 1 Split data into features and target variable
# Split the data into features and target label
cropyield = pd.DataFrame(df['Yield_per_ha'])

dropp=df[['Iron mg/kg','Copper mg/kg','Crop','Center','FarmID',
'Yield_per_ha']]
features= df.drop(dropp, axis=1)
cropyield.columns
features.columns
```

```python
#Step 2 Shuffle & Split Final Dataset
# Import train_test_split
from sklearn.cross_validation import train_test_split
from sklearn.utils import shuffle

# Shuffle and split the data into training and testing subsets
features=shuffle(features,  random_state=0)
cropyield=shuffle(cropyield,  random_state=0)
# Split the 'features' and 'income' data into training and
testing sets
X_train, X_test, y_train, y_test = train_test_split(features,
cropyield, test_size = 0.2, random_state = 0)

# Show the results of the split
print("Training set has {} samples.".format(X_train.shape[0]))
print("Testing set has {} samples.".format(X_test.shape[0]))

# Step 3 Model Building & Evaluation
#Creating the the Model for prediction

#Loading model Libraries
import matplotlib.pyplot as plt
import numpy as np
from sklearn import linear_model
from sklearn.metrics import mean_squared_error, r2_score

#Creating Linear Regression object
regr = linear_model.LinearRegression()

regr.fit(X_train,y_train)
y_pred= regr.predict(X_test)
#Printing Codfficients
print('Coefficients: \n',regr.coef_)
#print(LinearSVC().fit(X_train,y_train).coef_)
```

```
regr.score(X_train,y_train)
#Mean squared error
print("mean squared error:  %.2f" %mean_squared_error(y_test,
y_pred))
```

```
#Variance score
print("Variance score: %2f"  % r2_score(y_test, y_pred))
```

```
#Plot and visualize the Linear Regression plot
plt.plot(X_test, y_pred, linewidth=3)
plt.show()
```

```
#Predicting Yield per hectare for a new farmland
X_test.dtypes
predicted= regr.predict([[26,1500,6.8,0.9,367,32,490,35]])
print(predicted)
```

This is straightforward code. It first loads the dataset into a pandas dataframe and then checks data sanctity through df.size, df.dtypes, and other statements. The EDA is done with the df.mean(), df.median(), and df.mode() statements. The correlation is shown in Figure 9-1. In our case, we can have a look at the average mean values for soil pH and other soil nutrients in the output of Listing 9-2 and Figures 9-2 through 9-9. Outlier detection tells us that the yield data column has no outliers beyond the upper threshold limit; however, all the values are below the lower threshold limit. This can be seen visually by plotting the histogram using the command df.hist() in the code. After this, we can look at the correlation.

Correlation Vars	land size_in_ha	Crop_in_tonnes	Yield_per_ha	Soil_pH	Organic Carbon %	Nitrogen kg/ha	Phosphorus kg/ha	Potassium kg/ha	Iron mg/kg	Copper mg/kg	Sulphur kg/ha
landsize_in_ha	1										
Crop_in_tonnes	0.800644081	1									
Yield_per_ha	-0.612320185	-0.042910333	1								
Soil_pH	-0.661463232	-0.148609498	0.936215238	1							
Organic Carbon %	-0.539412303	-0.042020046	0.868255792	0.866357183	1						
Nitrogen kg/ha	-0.612562925	-0.167615712	0.831095999	0.812439244	0.724150383	1					
Phosphorus kg/ha	-0.564638105	-0.12045398	0.813284772	0.773350826	0.710242517	0.704619379	1				
Potassium kg/ha	-0.575397037	-0.150598611	0.783733627	0.767567692	0.720101262	0.711920492	0.653612933	1			
Iron mg/kg	0.115385734	0.103382496	0.030561931	0.062938634	0.317504032	0.080813021	0.012219977	0.05648151	1		
Copper mg/kg	0.014832656	0.018443531	0.006008786	0.012906811	-0.014122801	0.026012365	0.092460809	0.013622529	-0.076207204	1	
Sulphur kg/ha	-0.576157115	-0.149823618	0.792147864	0.785350631	0.729216708	0.667870193	0.665502786	0.693259322	0.040735434	0.026377934	1

Figure 9-1. *Correlation between the variables*

As we can see from Figure 9-1, we are interested in the Predictor Yield_per_ha or Yield per hectare. We think it is dependent on other variables or soil nutrients; this can be confirmed by looking at the correlation values between the Yield_per_ha variable with other variables. Soil_ph has a correlation of 0.936215238 with the Yield variable, Organic Carbon % has 0.868255792, Nitrogen kg/ha has 0.831095999, Phosphorus kg/ha has 0.831095999, Potassium kg/ha has 0.78733627, Iron mg/kg has -0.030561931, Copper mg/kg has 0.006008786, and Sulphur kg/ha has 0.792147864. To build our model, we pick the ones that have a significant correlation, namely

> Soil_pH
>
> Organic Carbon %
>
> Nitrogen kg/ha
>
> Phosphorus kg/ha
>
> Potassium kg/ha
>
> Sulphur kg/ha

We can ignore and remove iron and copper because they do not show any significant correlation; in fact, it is negligible for them to be considered for our model building exercise.

Listing 9-2. Output of Code from Listing 9-1

```
          Crop  Center  FarmID  landsize_in_ha
Crop_in_tonnes  Yield_per_ha  \
0  Sugarcane  Africa    1234              11.0
1235.3             112.3

    Soil_pH  Organic Carbon %  Nitrogen kg/ha
Phosphorus kg/ha  \
0      7.81              0.88            355.0
36.0
```

```
    Potassium kg/ha  Iron mg/kg  Copper mg/kg  Sulphur kg/ha
0              485.0        9.73          4.73  33.0
6314
(451, 14)
Index([u'Crop', u'Center', u'FarmID', u'landsize_in_ha',
u'Crop_in_tonnes', u'Yield_per_ha', u'Soil_pH', u'Organic
Carbon %', u'Nitrogen kg/ha', u'Phosphorus kg/ha', u'Potassium
kg/ha', u'Iron mg/kg', u'Copper mg/kg', u'Sulphur kg/ha'],
      dtype='object')
<class 'pandas.core.frame.DataFrame'>
RangeIndex: 451 entries, 0 to 450
Data columns (total 14 columns):
Crop                451 non-null object
Center              451 non-null object
FarmID              451 non-null int64
landsize_in_ha      451 non-null float64
Crop_in_tonnes      451 non-null float64
Yield_per_ha        451 non-null float64
Soil_pH             451 non-null float64
Organic Carbon %    451 non-null float64
Nitrogen kg/ha      451 non-null float64
Phosphorus kg/ha    451 non-null float64
Potassium kg/ha     451 non-null float64
Iron mg/kg          374 non-null float64
Copper mg/kg        337 non-null float64
Sulphur kg/ha       451 non-null float64
```

dtypes: float64(11), int64(1), object(2)

memory usage: 49.4+ KB

Training set has 360 samples.

Testing set has 91 samples.

('Coefficients: \n', array([[-2.11198278, 0.02322665,

5.78698466, 3.69351487, 0.01497725, 0.17939622, 0.01021635,

        0.15396713]]))

mean squared error: 16.50

Variance score: 0.950297

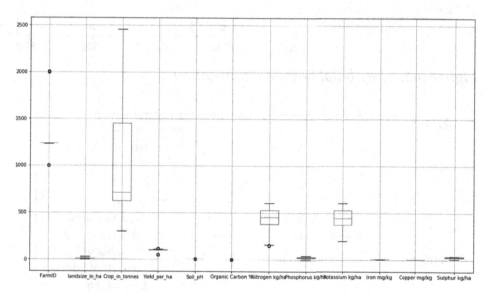

Figure 9-2. *Vertical boxplot visualization*

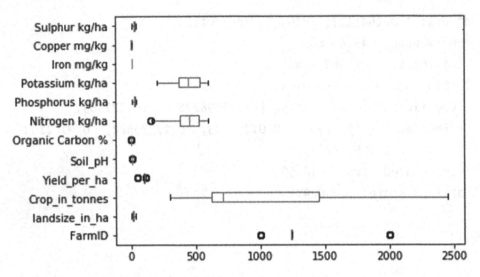

Figure 9-3. *Horizontal boxplot visualization*

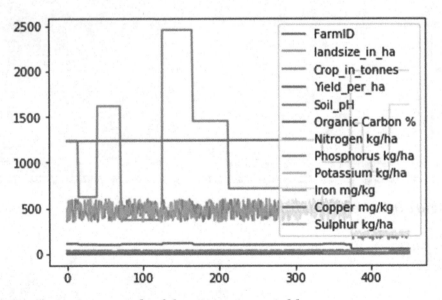

Figure 9-4. *Area graph of the numeric variables*

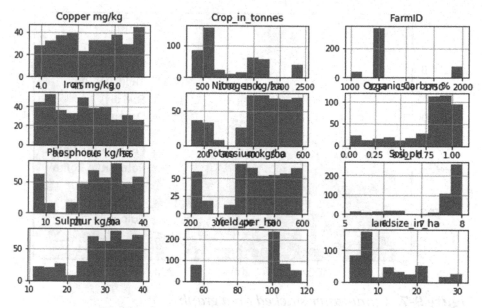

Figure 9-5. *Histogram of the numeric variables*

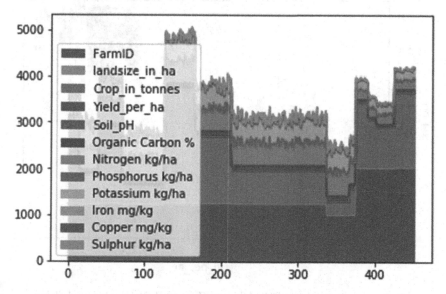

Figure 9-6. *Stacked area plot of the numeric variables*

267

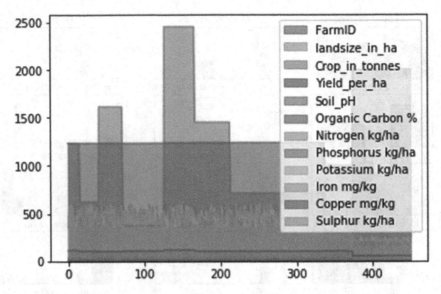

Figure 9-7. *Cumulative stacked area graph*

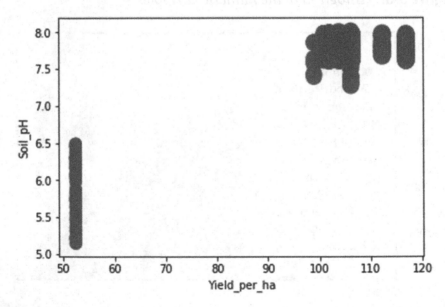

Figure 9-8. *Scatter plot of Yield_per_ha and Soil_oH variables*

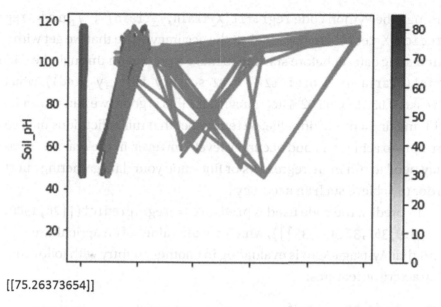

[[75.26373654]]

Figure 9-9. *Chart plot of Yield_per_ha and Soil_oH variables*

In Figures 9-2 through 9-8, we can see that the bloxplot shows us a highly distributed value of Crop_in_tonnes. One way to avoid seeing this is to apply scaling to all of the numerical variables, like I did in the case study solutions for healthcare, retail, and finance in my book *Machine Learning Applications Using Python*. The histogram shows the distribution of variables, and we see that most of the soil nutrients are right-skewed except for copper and iron. We also note that the scatter plots for Yield_per_ha and Soil_pH are closely related through the Python code `df.plot.hexbin(x='Yield_per_ha', y='Soil_pH', gridsize=20)`. After completing the usual data preparation steps by dividing it into a target variable named `cropyield` and a features variable with all the other features, we then split the dataset into training and testing datasets with 360 samples belonging to the training dataset and 91 samples belonging to the testing dataset. After this, we load the linear_model Python library to execute the linear regression algorithm on the training and testing datasets through the Python code `regr.fit(X_train,y_train)` , `y_pred= regr.predict(X_test)`. Then we look at the accuracy score that we get with our testing dataset before starting to make a prediction through the code `print("Variance score: %2f" % r2_score(y_test, y_pred))`, which gives us 0.9502, or 95.02% accuracy. Since this is good, we can proceed with making a prediction. Please remember that this is fictitious data so we are able to achieve a good accuracy level. However, in the real world, you may need to run more regressors or fine-tune your data gathering efforts in order to achieve such an accuracy level.

To predict, the code used is `predicted= regr.predict([[26,1500, 6.8,0.9,367,32,490,35]])`, which are the values of an agriculture land that Aystsaga Agro is evaluating in another country with following characteristic features:

```
landsize_in_ha       26
Crop_in_tonnes       1500
Soil_pH              6.8
Organic Carbon %     0.9
Nitrogen kg/ha       367
Phosphorus kg/ha     32
Potassium kg/ha      490
Sulphur kg/ha        35
```

Glanzo wants to know from the model what the probable yearly yield will be after they buy this farmland. The Python program gives out a value of 75.2637 tonnes per hectare, as shown in Listing 9-3.

Listing 9-3. Predicted Value by the Python Program

```
predicted= regr.predict([[26,1500,6.8,0.9,367,32,490,35]])
print(predicted)
[[75.26373654]]
```

Please remember that this program does not take into account other conditions such as weather and seed varieties, which also effect crop production and yield. Building such a dataset would definitely be a huge exercise beyond the scope of this book.

Summary

We have now come to the end of this case study and the book. I have thoroughly enjoyed bringing you these IoT-based solutions to modern-day, practical business problems and trying to solve them through machine learning. I hope you enjoyed learning from them too. Do consider leaving feedback on the forums at `www.pmauthor.com/raspbian`.

Index

© Puneet Mathur 2020
P. Mathur, *IoT Machine Learning Applications in Telecom, Energy, and Agriculture*,
https://doi.org/10.1007/978-1-4842-5549-0

Printed in the United States
By Bookmasters